Lecture Notes on Coastal and Estuarine Studies

20

Environmental Decisionmaking in a Transboundary Region

Edited by
Alison Rieser, Judith Spiller and David VanderZwaag

Springer-Verlag
Berlin Heidelberg New York London Paris Tokyo

Editors

Alison Rieser
University of Southern Maine, Marine Law Institute
246 Deering Avenue, Portland, Maine 04102, USA

Judith Spiller
University of New Hampshire, Complex Systems Research Center
Durham, New Hampshire 03824, USA

David VanderZwaag
Dalhousie University, Ocean Studies Program
Halifax, Nova Scotia, Canada B3H 4H8

ISBN 3-540-96446-0 Springer-Verlag Berlin Heidelberg New York
ISBN 0-387-96446-0 Springer-Verlag New York Berlin Heidelberg

Printed in Germany

Printing and binding: Beltz Offsetdruck, Hemsbach/Bergstr.
2131/3140-543210

PERSPECTIVE
SITE B-9

Preface

This monograph represents the work of an interdisciplinary, bilateral research team, with members from the Marine Law Institute of the University of Southern Maine, the Complex Systems Research Center of the University of New Hampshire, and the Ocean Studies Programme of Dalhousie University in Halifax, Nova Scotia. The study was made possible by grants from the William H. Donner Foundation, Inc. of New York and the Max Bell Foundation of Toronto, Canada. We are extremely grateful for their generous support and enthusiasm for the project.

In the course of the study the investigators held a number of workshops and meetings, allowing study participants to meet and exchange information among themselves and with representatives of government agencies, research organizations, and universities. In this manner additional expertise was brought to bear on specific facets of the study. Two of the most important meetings were held in the fall, 1983. The Canadian-U.S. Energy Exchange Workshop, co-sponsored by the Marine Law Institute, the Northeast-Midwest Institute (Washington, D.C.), and the Canadian-American Center of the University of Maine at Orono, was held October 24, 1983 in Portland, Maine to explore the potential for electricity imports and exports, the U.S. and Canadian regulatory frameworks controlling them, and Fundy tidal power's potential contribution to U.S. electricity requirements. On November 20-21, 1983, the Complex Systems Research Center sponsored a workshop on the regional scientific needs to address Fundy tidal power development. The agendas and lists of participants for these meetings are presented in the Appendices. We are grateful for the contribution made to this work by the workshop participants.

The investigators presented papers based on this study at a number of meetings, including the 1982 Fundy-Maine-Georges Bank (FMG) Forum in Digby, Nova Scotia, the 1982 annual meeting of the Coastal Society, the 1982 meeting of the New England Estuarine Society in Portland, Coastal Zone '83 Symposium in San Diego, the December, 1983 forum on Fundy tidal power sponsored by NOAA, Office of Coastal Resource Management, in Boston, and the 1983 Senate Committee on Environment and Public Works field hearing in Augusta, Maine. The response to these presentations helped to shape several of the final recommendations and, again, we are grateful for this assistance.

We wish also to acknowledge the assistance of Jill Cournoyer, research consultant to the Marine Law Institute, who prepared the subject index and assisted in proofreading; Ms. Judy G. Reade, Librarian, Dalhousie Ocean Studies Programme, who provided materials and reference assistance; Dr. Peter Ricketts, Assistant Professor of Geography, St. Mary's University in Halifax, and Ms. Julie Hutching, formerly a research assistant at the Institute for Resource and Environmental Studies, Dalhousie University, who prepared the Bay of Fundy - Gulf of Maine maps; and Dr. Arthur

Hanson, Director of the Institute for Resource and Environmental Studies at Dalhousie, and Dr. Douglas Johnston, Director of the Marine and Environmental Law Program and Professor of Law at Dalhousie, who served as senior project consultants. Thanks also to Ms. Linda Corkum and Ms. Ena Morris of the Dalhousie Ocean Studies Programme and Ms. Clara Kustra of the Complex Systems Research Center for typing and revising early drafts of the papers. The lion's share of our debt of gratitude is owed to Beverly Bayley-Smith, administrative coordinator of the Marine Law Institute, who retyped each chapter and prepared the final manuscript for typesetting, cheerfully coping with numerous revisions and technical problems under very tight deadlines.

Finally, we wish to thank Dr. Robert Goettel, Director of the Center for Research and Advanced Study, University of Southern Maine, for his assistance and encouragement through every phase of this endeavor.

Contributors

GORDON BEANLANDS. Dr. Beanlands is Director of Research for the Federal Environmental Assessment Review Office and Adjunct Professor of Environmental Studies at Dalhousie University. He holds B.Sc. and M.Sc. degrees in Forestry from the University of New Brunswick and a Ph.D. in Ecology from Dalhousie University. He is the author of various articles on land use and coastal zone management and is co-author of *An Ecological Framework for Environmental Impact Assessment in Canada*. Institute for Resource and Environmental Studies, 1312 Robie Street, Halifax, Nova Scotia, Canada B3H 3E2.

PETER DUINKER. Mr. Duinker was a Research Associate at the Institute for Resource and Environmental Studies, Dalhousie University, from 1981-1983 and is presently completing a Ph.D. program in Forestry at the University of New Brunswick. He holds the B.Sc. Agr. degree from Guelph University and an M.E.S. degree from Dalhousie University. He is the author of many journal articles on environmental impact assessment and is co-author of *An Ecological Framework for Environmental Impact Assessment in Canada*. Faculty of Forestry, University of New Brunswick, Fredericton, New Brunswick, Canada E3T 6C2.

CYNTHIA LAMSON. Ms. Lamson is a Research Associate with the Dalhousie Ocean Studies Programme and a Lecturer-in-Law at Dalhousie Law School. She holds degrees from Simmons College (M.S. Library Science), Memorial University (M.A. Folk Lore), and Dalhousie University (M.E.S.). She is author of a book on the Canadian sealing controversy, *Bloody Decks and a Bumper Crop*, and co-editor of *Atlantic Fisheries and Coastal Communities: Fisheries Decision-Making Case Studies.* Dalhousie Ocean Studies Programme, Central Services Building, 5th Floor, 1236 Henry Street, Halifax, Nova Scotia, Canada B3H 3J5.

KAREN A. MASSEY. Ms. Massey was a Research Assistant and Staff Attorney with the Marine Law Institute, University of Southern Maine, until April, 1985, and is now Secretary of the Land and Water Resources Council, Maine State Planning Office. Her academic background includes a bachelor's degree from Middlebury College and a law degree from Yale Law School. Maine State Planning Office, Augusta, Maine 04333.

ALISON RIESER. Ms. Rieser is Director of the Marine Law Institute, University of Southern Maine, and Lecturer at the University of Maine School of Law. She was formerly a Research Fellow at the Woods Hole Oceanographic Institution and an Attorney-Advisor for Fisheries with the National Oceanic and Atmospheric Administration, Office of General Counsel. Her undergraduate degree is from Cornell

University and her law degree frm the National Law Center of the George Washington University. She is editor of TERRITORIAL SEA, a quarterly publication on legal developments in the management of interjurisdictional marine resources, and the author of several papers and articles on state-federal relations in fishery regulation and marine environmental policy. Marine Law Institute, 246 Deering Avenue, Portland, Maine 04102.

JOHN ROANOWICZ. Mr. Roanowicz was a Research Assistant with the Complex Systems Research Center, University of New Hampshire, from 1982-1983. He holds a bachelor's degree from Harvard University and a Masters of Marine Affairs from the University of Rhode Island. Evans Road, Madbury, New Hampshire 03824.

JUDITH SPILLER. Dr. Spiller is an Assistant Research Professor at the Complex Systems Research Center at the University of New Hampshire. She was formerly a Policy Associate at the Woods Hole Oceanographic Institution. Her bachelor's degree is from Columbia University and her Ph.D. from the State University of New York at Stony Brook. She is the author of numerous papers on the use of science in environmental assessment and regulation. Complex Systems Research Center, O'Kane House, University of New Hampshire, Durham, New Hampshire 03824.

ROBERT A. TAYLOR. Mr. Taylor was a Research Assistant and Staff Attorney with the Marine Law Institute from 1982 to 1983. He is now an attorney with the NOAA Office of General Counsel for Enforcement and Litigation. His undergraduate degree is from the University of Alabama, his J.D. from the University of Oregon, and L.L.M. in Marine Affairs from the University of Washington. NOAA, Office of General Counsel, 3300 Whitehaven Street, N.W., Washington, D.C. 20235.

PETER UNDERWOOD. Mr. Underwood was a Research Associate with Dalhousie Ocean Studies Programme until February, 1985, and is now with the Department of Fisheries of the Government of Nova Scotia. He holds a law degree from Dalhousie Law School and an M.Sc. degree in oceanography from the State University of New York. He has contributed to various books related to ocean management including *The Implementation of the New Law of the Sea in West Africa* and *Management and Utilization of the Marine Resources of the British Virgin Islands*. Nova Scotia Department of Fisheries, Purdy's Wharf Tower, P.O. Box 2223, Halifax, Nova Scotia, Canada B3H 3J5.

DAVID VANDERZWAAG. Mr. VanderZwaag is a Research Associate with the Dalhousie Ocean Studies Programme and a Lecturer-in-Law at Dalhousie Law School. His educational background includes a M.D.D. from Princeton Theological Seminary, a J.D. from the University of Arkansas School of Law, and the L.L.M. in international law of the sea from Dalhousie Law School. He has published various articles related to ocean management issues and is author of *The Fish Feud: The U.S. and Canadian Boundary Dispute*. Dalhousie Ocean Studies Programme, Central Services Building, 5th Floor, 1236 Henry Street, Halifax, Nova Scotia, Canada B3J 3J5.

List of Figures

Contents

Introduction

Alison Rieser

An increasing number of energy-related projects are being constructed or considered close to the United States - Canadian border, several of them designed chiefly to produce power for export across the boundary. Such projects include the Point Lepreau nuclear power station in New Brunswick, the offshore oil and gas exploration in the Beaufort Sea, tidal power barrages in the upper Bay of Fundy, and the High Ross Dam hydro-electric project in Washington State near the British Columbia border. (Figure 1 illustrates the number of energy projects in the Gulf of Maine region that were under consideration in 1984.) Like most large-scale energy developments, these proposals raise significant environmental issues.

Each of these developments occurs within the territory of one country, but its proximity to the international border makes it a matter of bilateral concern. How environmental considerations are factored into decisions to proceed with such projects depends upon the environmental policies, laws, and assessment procedures of the country in which the project is to be carried out. The transboundary nature of some of the environmental impacts, however, are likely to raise international political considerations that can obscure the technical and scientific questions presented by the potential impacts. Even a comprehensive national assessment process can be derailed by the widespread misunderstanding that is fostered by the surrounding controversy.

As many of the recent, large-scale energy developments in Canada rely on export markets in the United States, the U.S. and Canada are increasing their energy interdependence. Because such partnerships are likely to continue, and many will entail transboundary, environmental consequences that are difficult to predict and control, it is necessary and timely to consider whether the frameworks for environmental decisionmaking in the two countries are capable of addressing transboundary effects adequately. If there are inherent limitations in national systems for addressing transborder effects, one must consider whether bilateral institutions may be better suited for resolving such concerns. It is in both nations' interests to recognize the mutual benefits of both energy development and environmental protection. If projects with transboundary environmental impacts can be jointly considered and found to be mutually acceptable after a full consideration of all available data and information, then disruptive bilateral conflict over shared environmental resources is avoided and worthwhile energy development is promoted.

As a way of structuring an examination of U.S. and Canadian environmental decisionmaking on projects affecting a transboundary region, this work examines the Bay of Fundy tidal power project proposed for Nova Scotia, a long-standing proposal revitalized lately by changing internationl energy conditions. The proposed project

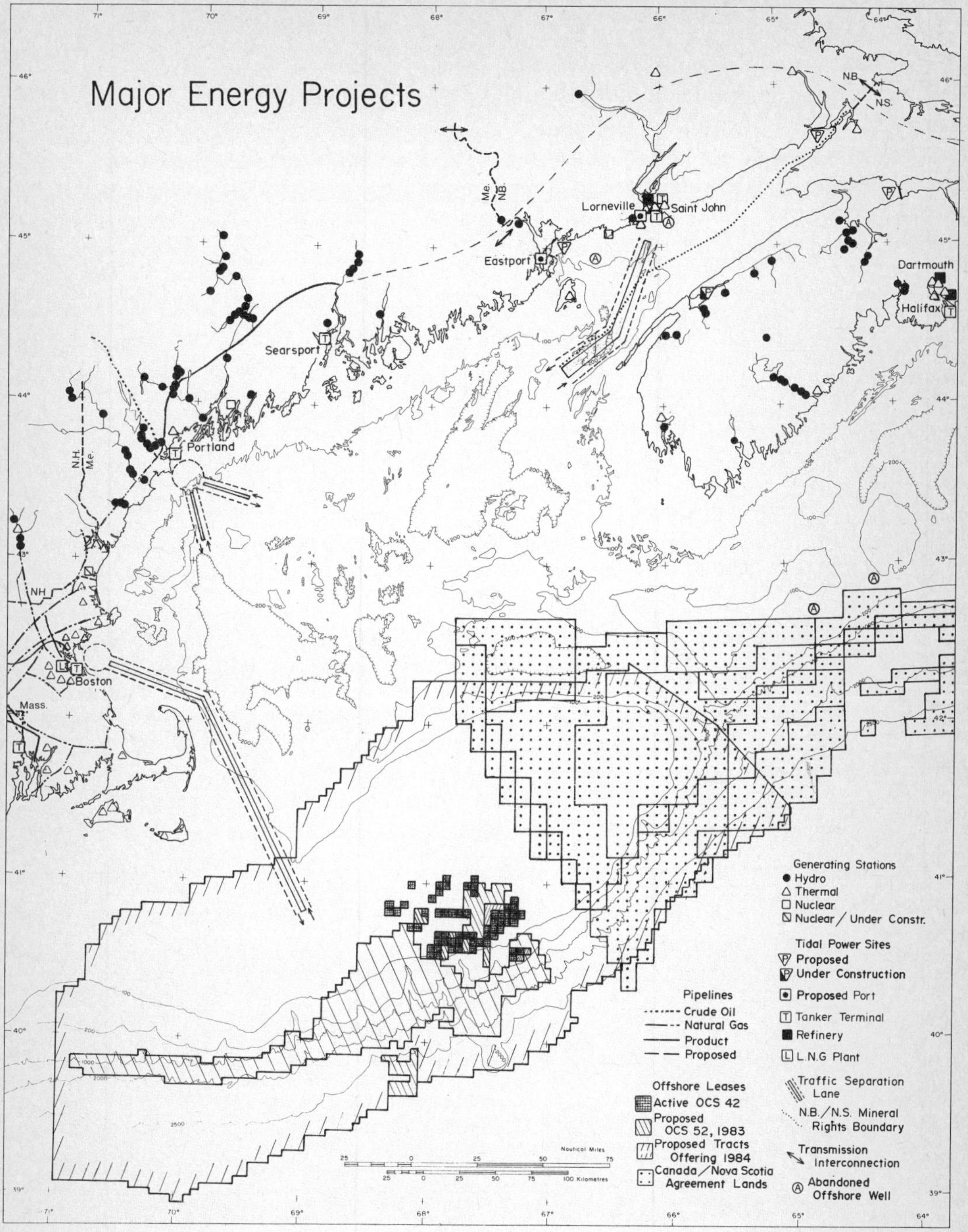

Figure 1. Major Energy Projects in the Gulf of Maine Region.

Map prepared by P.J. Ricketts, Saint Mary's University and J. Hutchings, Dalhousie University. Published with permission.

would harness the enormous tidal energy of the Bay of Fundy through the construction of large tidal barriers, up to five miles in length, equipped with turbines capable of generating up to 5,000 megawatts of electricity. The major portion of this power would be exported to markets in New England and New York. Tidal power is of particular interest because it involves an alternative, renewable, and large-scale energy source. Due to the nature of the Gulf of Maine-Bay of Fundy tidal resource, preliminary scientific studies predict the project would cause an increase in tidal levels along the New England coast, with implications for shorefront property and structures, commercial fisheries, and wetland functions.

Despite the long history of bilateral cooperation, environmental relations between Canada and the United States are particularly strained at present, primarily due to disagreements over actions necessary to address acid precipitation, one of the international community's most pressing transboundary environmental issues. In the New England-Atlantic Maritimes region, the maritime boundary for the Gulf of Maine and Georges Bank, settled in October, 1984 by the World Court, has increased rather than diminished the potential for conflict over shared resources.

The purpose of this study is to provide a basis for the cooperative consideration of developments that present benefits and costs to both sides of the border, in recognition that geography and history make us neighbors and that our shared resources cannot be managed effectively without such cooperation.

Chapter 1 looks at the Nova Scotian tidal power projects as case studies of technology assessment and illlustrates one of the most important elements of environmental assessment practice: the acquisition and sharing of technical information among project proponents, outside experts, and citizen groups. The chapter also provides important chronological information about the assessments of Fundy power that have been conducted to date.

Chapter 2 reviews the major scientific questions which surround the assessment of Fundy tidal power's environmental impacts. It describes the tidal system in the Bay of Fundy - Gulf of Maine and the variability of the forces which structure the ecosystem. It also examine the computer model that is the basis for the predicted alterations to the tidal regime and the resulting secondary effects. It outlines some of the scientific work that must be done to provide a better assessment of the potential changes, concluding that any research must recognize that the region functions as a system.

Chapter 3 provides a general overview of the Canadian and U.S. laws that would play an important role in shaping the decisions concerning Fundy tidal power. The chapter's purpose is to identify the review procedures and substantive criteria that Canadian and U.S. law and practice would impose on Fundy tidal power and to compare their major features. To illustrate the U.S. approach and to thereby provide a basis for comparison, the federal and state laws potentially applicable to a large-scale, hydro-electric project in the U.S. are also examined. The analysis finds certain similarities in the two systems in the overlap of authority among federal and state/provincial agencies, in the potential for interagency disagreement, and in the opportunity for judicial review of agency decisions. While the U.S. system is more highly structured because of the numerous and explicit statutory requirements, it may not guarantee a more thorough review of the transboundary environmental issues.

Chapter 4 examines in detail the environmental assessment procedures of the two countries to determine their effectiveness in the consideration of indirect and transboundary environmental effects. In its comparison of the two systems, it presents a recent case study of U.S. environmental decisionmaking -- the formerly proposed oil refinery and marine terminal at Eastport, Maine -- and concludes that political and institutional problems can prevent the effective consideration of transboundary effects. Despite a number of differences, however, both the U.S. and Canadian systems have the potential to address adequately and fairly the kind of impacts that are associated with the Fundy tidal power proposal.

Chapter 5 considers the remedies that might be available through the courts to citizens who are adversely affected by a project like Fundy tidal power. It also reviews the international law principles that specify the rights and responsibilities of countries in the conduct of activities with transboundary environmental effects.

Chapter 6 makes specific recommendations to address the major deficiencies present in the Canadian envionmental assessment process. It also describes ways in which the U.S. legal framework could be amended to trigger an environmental assessment of Fundy tidal power under U.S. laws. Recognizing that joint action could be more effective than unilateral action, even under strengthened national and state/provincial laws, the chapter reviews a number of bilateral arrangements, including informal consultations, a reference to the International Joint Commission, creation of a new bilateral organization, and adoption of a regional action plan for the shared marine environment.

A summary of the major conclusions and recommendations is presented in chapter 7.

Chapter 1

Technology Assessment and Nova Scotian Tidal Power Projects

Cynthia Lamson

Introduction

Proposals to construct and operate large-scale energy projects, particularly coastal and marine energy projects, tend to be controversial and highly politicized. Because technological change outpaces our understanding of the possible effects a new technology may have on the environment, and because scientific understanding about ecosystem dynamics is incomplete, development decisions are often based on limited information and are made in a climate of uncertainty. Decisions to proceed with (or delay) a project are, therefore, frequently the product of consensus-seeking with respect to the question: what constitutes politically acceptable levels of risk? "Until recently, evaluating the risks of technology has been considered a technical problem, not a political issue, a problem to be relegated to expertise, not to public debate. But controversies have increasingly politicized the issue of risk."[1]

Controversies arise because members of the engineering and scientific communities often disagree among themselves about the acceptability of predicted risks, and because special interest groups compete with one another to persuade decisionmakers that benefits of development outweigh the costs (or vice versa). Furthermore, acknowledgement that scientific uncertainty exists often has a "carry over" effect so that the focus of debate shifts from assessing environmental risk to critiquing decision-making procedures: "concerns about risk are translated into questions about the legitimacy of decisionmaking procedures and the credibility of scientific expertise as a reliable basis for them."[2]

In the quest to defuse controversy and confrontation, decisionmakers often rely on advisory and consultative mechanisms such as special committees, studies, and task forces to assist them in making politically acceptable decisions. Some of these mechanisms are exclusive and elite, that is, "expertise" is carefully selected and access to information is controlled; other mechanisms are more open and participatory by design, that is, input and opinion is solicited from a wide variety of interest groups.[3] Taken together, all of the procedures and mechanisms which comprise a "decision system" are the product of perceptions about a particular issue or project, and reflect how decisionmakers approach the tasks of problem identification and evaluation.

This chapter examines these aspects of technology assessment by providing a close-up picture of the Canadian environmental decisionmaking system through two case studies: the Annapolis River (Nova Scotia) tidal power facility and the proposed Fundy tidal power project. Emphasis is on questions related to process, participation, issue identification, and information management. This approach is adapted from technology assessment practice which focuses on "who" and "how" questions such as: (a) whose interests are central to project promotion or rejection; (b) what procedures and mechanisms are used to evaluate project proposals; and (c) how, and by what criteria, decisions are ultimately made.[4]

These case studies illustrate how a legally sanctioned assessment system shapes and controls assessment practice. It demonstrates how information and data may be managed by powerful interest groups to promote particular points of view, and how decision systems may be susceptible to political pressure and manipulation as a by-product of their structural flexibility.

Fundy Tidal Power Proposals

The dream of harnessing power from the tides of the Bay of Fundy - Gulf of Maine region is a recurring phenomenon. Since the 1920's, a variety of projects have been promoted and numerous feasibility studies have been conducted on both sides of the Canadian - U.S. border.[5] Canadian interests have been seriously evaluating the feasibility of a tidal power facility in the upper reaches of the Bay of Fundy since August, 1966, the date of the first Federal/Provincial Tidal Power Study Agreement. In a sense, the Fundy case study is still evolving since no firm decisions have been made to proceed with the development, nor has a final site been selected. Large-scale development of tidal power has not been subjected to formal review by the Environmental Assessment and Review Process (EARP) or the National Energy Board (NEB). Therefore, the Fundy project must be examined in the context of decisions and assessment activity which has taken place to date, and it may be more accurate to describe the case study as an analysis of *assessment practice* rather than decisionmaking.

During the protracted review period, several important questions have emerged which will be central to future decisionmaking about Fundy tidal power. These questions are examined in detail in chapter 2. They are: a) the range and extent of impacts that tidal power generation may have on the physical oceanographic regime in the Bay of Fundy - Gulf of Maine region, and b) the nature and extent of interactive, or ecological systems effects that may occur as a result of harnessing tidal power. These questions are, however, a subset of socio-economic and political considerations which decisionmakers will be forced to address if a serious development proposal is brought forward. Specifically, the issues raised are:

1. the ability to predict the magnitude of change and the extent of disruption that will be experienced by resource users if a large-scale tidal power generating facility is constructed;

2. the ability to mitigate adverse impacts;

3. the acceptability of proposed compensation schemes;

4. finding appropriate mechanisms to resolve transboundary conflict; and

5. achieving political consensus across provincial, federal, and national boundaries that the benefits of constructing a large-scale tidal power facility outweigh the costs of pursuing alternative energy sources.

Documentation of tidal power decisionmaking is somewhat problematic because most of the assessment activity to date has focused on the evaluation of a *concept*, that is, the feasibility of power generation from the tides of the Bay of Fundy. The case study is complicated further by the fact that, in the midst of debate about a large-scale facility, a pilot tidal power project at Annapolis Royal was proposed, reviewed, and approved. Construction of the Annapolis Royal site started in May, 1980, and the facility commenced operation in May, 1984.

With the Annapolis development, Fundy decisionmaking must be viewed as a two-part, albeit unfinished, story. Although questions of economic feasibility have been of paramount concern to decisionmakers, in the course of the past seventeen years a number of issues have been raised which affect the estimation of economic feasibility. Specifically, these are the costs associated with: (i) environmental and ecosystem disruption or alteration, (ii) mitigation, and (iii) compensation. Although the assessment record reveals that only peripheral attention has been paid to these questions to date, proponents will be expected to demonstrate some degree of concern (through provision of study data and planning reports) when a specific project proposal is submitted for review by EARP and the NEB.

The Fundy case study documents the evolution of assessment in terms of how project promoters, reviewers, and evaluators, as well as the public have:

1. *perceived* potential environmental benefits and risks associated with tidal power development;

2. *assessed* potential impacts; and

3. managed information to focus debate on certain issues to the exclusion or postponement of other questions.

The various investigations of the feasibility of Fundy tidal power are discussed in chronological sequence.

Early Feasibility Studies

Under the terms of a 1966 Federal/Provincial Tidal Power Study Agreement, $1.5 million was allocated to the Atlantic Tidal Power Programming Board (ATPPB or

Board) and the Atlantic Tidal Power Engineering and Management Committee to: (a) review previous studies of tidal power sites, and (b) produce a preliminary engineering design, as well as preliminary marketing and transmission studies. The Board was instructed: "In carrying out these studies, the Board will assess the effects of the power schemes on other interests such as fisheries, navigation and transportation as well as upon the tidal power regime of the Bay of Fundy."[6]

A Navigation Sub-Committee reported that development at Site 8.1 [B9] (Cobequid Bay/Economy Point) (Figure 2) would involve slight changes in tidal regime (lowered high-water levels) and that dredging would be necessary in Hantsport, Walton, and Port Williams if current traffic patterns were maintained. A Fisheries Sub-Committee concluded that development would not adversely affect the commercial lobster, herring or groundfish fisheries in the Upper Bay of Fundy, although some disruption of the soft-shell clam fishery was predicted if the Economy Point site was developed. Although the Board's report cautioned that incomplete biological knowledge and difficulties associated with forecasting environmental changes made precise predictions impossible, it did not perceive that any additional surveys or studies of the fisheries were necessary at the present time. In brief, the Engineering and Management Committee advised the ATPPB that it perceived impacts on agriculture, forestry, ground transportation, and navigation would be minimal:

> The effects of a tidal power project on agriculture, forestry, fisheries, ground transportation and navigation were briefly investigated and, in general, were found to be minimal. Low-cost power could have a favourable impact on the region's agriculture development by reducing the cost of frozen-fruit and vegetable processing. Forestry is expected to benefit from lower transportation costs through better project-related access roads including the roads over the structures themselves. The adoption of Site 7.1 would reduce the highway distance from St. John to Halifax by five miles and the use of Site 8.1 by 15 miles.[7]

Despite these conclusions, the Committee reported to the ATPPB in October, 1969, advising that "on the basis of a seven percent interest rate, none of the tidal power schemes investigated would be competitive with alternative sources of generation."[8] The Board, in turn, accepted this assessment and recommended that additional studies should be authorized *only* when economic factors improved the competitiveness of tidal power, or when alternative energy sources became exhausted.

The findings and recommendations of the ATPPB Report, presented to Parliament on March 20, 1970, disappointed many interests, particularly in Nova Scotia. Almost immediately, the Report and the Committee became the focus of political controversy. Some observers suggested that hydro-development interests in Quebec feared competition from tidal power development in the Maritimes and were responsible for delaying the Fundy project.[9] The ATPPB Report was also criticized for relying on outdated, conventional economics, and the failure to calculate pollution costs associated with existing energy production modes was perceived by Fundy advocates as a major oversight in terms of comparative analysis.[10]

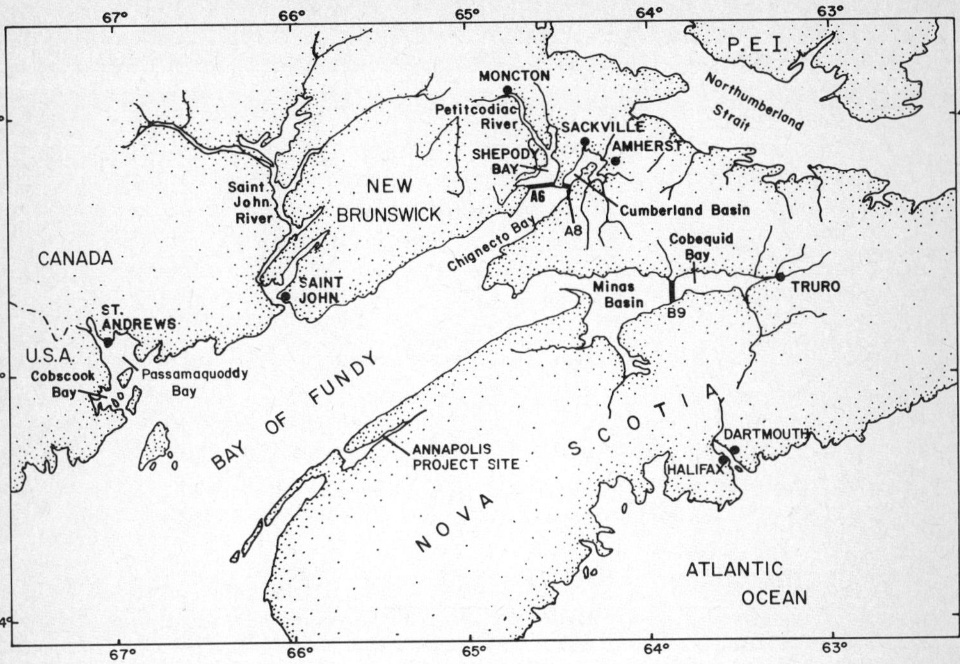

Figure 2. Proposed Bay of Fundy Tidal Power Sites.

Adapted from Bay of Fundy Tidal Power Review Board, *Reassessment of Fundy Tidal Power* (1977), p. 15.

After a year-long debate, Fundy tidal power development remained in doubt. Then, in late March, 1971, the proposal regained center stage in both federal and provincial political circles. First, Nova Scotia Premier Gerald Regan announced the creation of the Tidal Power Corporation, a provincial Crown corporation (Tidal Power Corporation Act, N.S., 1970-71, c. 21). This unilateral action on the part of the Province baffled a number of tidal power proponents who believed that Nova Scotia alone could not afford the estimated $1 billion for the Fundy tidal project, and that participation by Ottawa and New Brunswick was essential.[11] The following week the federal Energy Minister, J.J. Greene, established an all-party parliamentary committee, chaired by Nova Scotia Member of Parliament Robert Coates, to investigate the possibility of conducting a reassessment of the tidal power energy option.[12]

Reassessment of Fundy Tidal Power, 1972-1977

On February 29, 1972, the Governments of Canada, New Brunswick, and Nova Scotia established the Bay of Fundy Tidal Power Review Board (BFTPRB) for the purpose of reviewing the findings of the ATPPB's 1969 Report in light of current and projected fiscal circumstances. The Review Board reported in September, 1974, concluding that the relative economic position of tidal power had improved, and that a more comprehensive assessment was warranted. A number of studies were identified as being critical to project approval, including: (i) domestic and export markets; (ii) transmission; (iii) retiming; (iv) system studies; and (v) environmental impact. However, with reference to environmental studies, the Board reiterated (and accepted) the conclusions of the 1969 ATPPB assessment, that is, that environmental effects would not be anticipated to alter a decision on the economic merits of development, and stated:

> An environmental reappraisal for the purposes of this preliminary assessment of the economic viability of a tidal power development was not warranted. However, further intensive engineering and economic studies should embrace environmental aspects. The Review Board has had developed by Environment Canada the framework for the type of study necessary to identify beneficial or adverse effects on the environment (and, possibly to quantify any such effects) that would likely result from construction and operation of a tidal power development and required associated facilities.[13]

In response to the 1974 BFTPRB Report, a third federal-provincial agreement was signed in December, 1975, for the purpose of "providing a firm estimate of the cost of tidal energy in relation to its alternatives, on which to base a decision to proceed further with detailed investigations and engineering design."[14] The agreement specified that a seven-member Management Committee would be responsible for conducting a study program in five task areas: (1) tidal power plant design; (2) generation optimization; (3) markets and systems analysis, alternative supplies and transmission; (4) socio-economic aspects; and (5) environmental aspects.

The Management Committee Report was released in November, 1977, but only twenty-one pages of the 516-page document addressed socio-economic and environmental concerns associated with tidal power development. The Committee stressed that its findings were preliminary, largely qualitative, and based on the present state of knowledge: "evaluations have been restricted to an initial exploration of potentially significant issues based on *existing information* and *experience elsewhere*."[15] Three categories of environmental impacts were identified as requiring attention in subsequent assessment studies including: environmental alteration, effects on natural processes, and natural resources availability. In addition, the Committee determined a tentative priority ranking of perceived environmental issues:[16]

HIGH: agriculture, drainage and flooding

MODERATE: birds and anadromous fish

LOW: other fisheries, recreation, minerals, and navigation

Based on previous research effort, the Committee noted that modifications of the tidal regime would place certain marshland habitats and certain fish and bird populations at risk:

> While the loss of mudflat and marshland resources could be significant, it has been acknowledged that land reclamation has had a greater impact on avifauna than is likely to result from a tidal power development. Experience elsewhere has also shown that shorebirds are seldom food-limited. They will crowd together on mudflats in response to the loss of habitat and show considerable flexibility in habitat selection.[17]

Fisheries concerns were classified under three headings: passage of fish through a barrage, production in headpond or tidal basin areas, and effects on fisheries in the outer Bay. Although limited information was noted, the Report advised: "Concerns have been expressed but may not prove to be serious when additional surveys have been conducted."[18] The Committee considered possible effects of regime alteration on selected anadromous species, but failed to discuss the potential impacts of habitat alteration for the commercial and recreational fisheries, except for a passing reference to the clam fishery: "the extensive clam fishery in the Economy Point-Five Island area and the recreational fishery in the Minas Basin should not be unduly affected by tidal development."[19]

Potential changes in drainage patterns and groundwater levels were recognized by the Committee as posing some degree of threat to existing high quality pasture and prime farmland. The Committee also expressed concern that slight alterations in local air temperatures might have adverse effects on agricultural practices. Given that two-thirds of the Province's annual horticultural production value derives from Kings, Hants, and parts of Colchester and Cumberland Counties, the Report noted: "any reduction in degree-days or the frost-free period could be detrimental to agricultural production in these areas."[20] Recreational and tourism concerns were briefly addressed in the Report. Although increases in overall utilization of the area were anticipated to follow development, one negative aspect was cited: "changes in tidal amplitude may affect beach erosion in shore areas requiring monitoring and subsequent remedial action, while habitat modifications could affect established recreational patterns."[21] Mineral resource extraction activities (coal and salt) were perceived to be vulnerable if elevations in groundwater tables occurred, but the question was not ranked as a priority issue for subsequent study. In contrast to the 1969 ATPPB Report, the 1977 Reassessment did not forecast the need for harbor dredging to facilitate navigation or other marine operations.

The BFTPRB concluded that tidal power was potentially feasible in economic terms and advocated the undertaking of more detailed studies at the Cumberland Basin (A8) site, and possibly at Economy Point in the Minas Basin (B9) and Shepody Bay (A6). The principal (perceived) constraint to development was the resolution of financial participation, while potential environmental impacts or ecosystem alterations were not judged as being significant factors to the project decision process: "Environmental considerations are unlikely to be of sufficient importance to affect a decision to proceed with development of a tidal power project or to result in a strong preference for any site."[22] Table 1 summarizes environmental risks identified by the ATPPB (1969) and the BFTPRB (1977) reports.

Current Status of Proposal

Tidal power -- once actively investigated by the New Brunswick and Nova Scotia, as well as the federal, governments -- is now almost exclusively a Nova Scotian project. New Brunswick dropped out in the mid-seventies when its government gave priority to nuclear energy. Consequently, development proposals have shifted from Site A8 (Cumberland Basin, between New Brunswick and Nova Scotia) towards B9 (Minas Basin). Although the 1977 BFTPRB Reassessment favored Site A8 for economic reasons, the Tidal Power Corporation now promotes B9 (entirely within Nova Scotia boundaries), arguing that a larger facility generating 4,864 megawatts (mgw) was preferable to a 1000 mgw project. To substantiate this position, the Tidal Power Corporation recalculated cost projections for each site and was able to demonstrate improved benefit/cost ratios for the B9 Site as compared with Site A8.[23]

Progress towards development of a large-scale project has been stalled while financing for feasibility studies and commitments from potential buyers are being sought. In May, 1984, the federal and provincial Ministers of Energy jointly announced that financing for an updated feasibility and marketability study was forthcoming, and the study would require approximately nine months to complete.[24] Alcan Aluminum Ltd. was commissioned in 1985 to act as a 'catalyst' for a major project, and undertook the task of evaluating capital costs, potential revenue, market and site possibilities, and costs for generation and transmission.[25] A formal environmental assessment and review process inquiry will not be conducted until financing and marketing issues are resolved, and it is certain that the draft environmental impact statement (EIS) guidelines (released April, 1978) will be revised to address project-specific design, construction, and operational questions.[26]

Table 1

Risk Identification Matrix: Large-Scale Development[1]

	Identified Risks	Resources/Activities at Risk: SHIPPING	AGRICULTURE	FISHERIES	BIRDS	RECREATION
ATPPB (1969)	changes in mean water levels	●	●	●	●	
	changes in turbulence			●		
	changes in sedimentation	●	●	●	●	●
	reduction of intertidal zone				●	●
	changes in biological activity			●		
	increased erosion		●			
	changes in drainage patterns		●			
BFTPRB	impaired drainage		●			
	raised groundwater levels		●			
	climatic changes		●			
	river estuarine flooding		●			
	increased ice cover		●	●	●	
	changes in sedimentation patterns (transport and deposition)		●	●	●	
	reduction of mudflat and marshland resources			●	●	●
	barrage construction	●		●		●
	changes in biological productivity			●	●	
	changes in tidal amplitude		●			●

Note: [1] Environmental risks identified by two 'feasibility assessment' documents, i.e., Feasibility of Tidal Power Development in the Bay of Fundy (ATPPB, 1969), and Reassessment of Fundy Tidal Power (BFTPRB, 1977).

Interest in development of a large-scale tidal power project in the Bay of Fundy has been sustained through a variety of mechanisms including scientific forums, government-sponsored symposia and hearings, and by the media. As a result, public interest in Fundy tidal power has broadened and new concerns about the potential impacts of developments have been raised. Thus, assessment activity is no longer restricted to technological or economic feasibility questions, and studies examining some of the potential environmental impacts of tidal power development are underway. In Canada, environmental studies in the Bay of Fundy were conducted initially on a voluntary and *ad hoc* basis. Cooperative activity and research on a regional basis traces back to the convening of a small workshop at Acadia University in 1976 and the formation of the Fundy Environmental Studies Committee. In contrast, research activity by U.S. scientists and agencies was mobilized in response to political mandates which were, in turn, responses to public opinion and interest group lobbying.

Fundy Scientific Studies

In November, 1976, the Management Committee of the BFTPRB, in conjunction with the Acadia University Institute and Environment Canada, convened a workshop on the "Environmental Implications of Fundy Tidal Power" at Acadia University (Wolfville, N.S.). Under the chairmanship of G.C. Baker, the workshop was designed to meet four objectives:

1. to provide a forum for the dissemination of information on possible Fundy tidal power developments, their location and probable environmental effects;

2. to conduct a review of the existing knowledge on the Bay of Fundy, Minas Basin and Gulf of Maine systems;

3. to initiate communications between scientists of differing disciplines with a common interest in the Bay of Fundy/Gulf of Maine system in order that the future environmental studies shall be carried out with regard for interdisciplinary considerations;

4. to provide the Fundy Tidal Power Management Committee with advice on the design and commitment of future studies into environmental effects.[27]

The workshop provided a forum for scientists to share their research with colleagues, and served as a catalyst for cooperative activity and communications. A second workshop was convened in 1977, and an *ad hoc* committee, the Fundy Environmental Studies Committee (FESC), was created at that second gathering.

The FESC is comprised of about 150 individuals who represent organizations that conduct or support research or development activities in the Bay of Fundy. For the first few years, FESC was an informal association of scientists and activities were administered from the Marine Ecology Laboratory, Department of Fisheries and Oceans (DFO), located at the Bedford Institute of Oceanography (Dartmouth, Nova Scotia). In 1979, the group was formally organized as a Committee under the aegis of the Atlantic Provinces Council on the Sciences (APCS), located in Halifax.

FESC may be said to embody a "free enterprise" approach to research, whereby individual scientists are permitted to pursue their own research interests and programs, but are encouraged to coordinate their work with others pursuing related research in the same area. Despite concerted efforts to secure direct involvement of U.S. researchers with the FESC, participation to date has generally been low. Consequently, the emphasis of the ecological studies coordinated by FESC has been on understanding the effects of tidal power development on Canadian resources. APCS has sponsored annual public meetings of FESC since 1977, and many members have gained recognition as Canadian experts on the potential environmental effects of tidal power development.[28]

A number of other scientific forums have been convened in recent years with both direct and indirect linkages to the Fundy Environmental Studies Committee. For example, a Symposium on the Dynamics of Turbid Coastal Environments (colloquially called the "Muddy Waters Symposium") was convened at the Bedford Institute of Oceanography in September, 1981, and a second meeting was held in the United Kingdom in 1985, sponsored by the Estuarine and Brackish Water Sciences Association. In March, 1983, FESC members attended a workshop in the Netherlands to consider the feasibility of adapting the EMS Dollard estuary model to the Cumberland Basin. Canadian scientists convened follow-up workshops in 1984 at Acadia University and at the Bedford Institute of Oceanography. Meanwhile, estuarine models were also being developed at the University of New Hampshire and the University of Maine. In November, 1983, a small workshop was convened at the University of New Hamsphire to address "Regional Scientific Needs in Fundy Tidal Power Development." Scientists from the U.S. and Canada attempted to evaluate the reliability of existing predictions about the effect of Fundy tidal power development and to prioritize future directions for research. Participants were in general agreement that model predictions were accepted as being very good; however, three sources of potential error were identified: resolution of depths in areas with rough topography, the accuracy of boundary conditions, and the computation of tidal energy dissipation.[29]

U.S. Responses to Proposed Fundy Tidal Power Project

Although U.S. scientists did not attend early meetings of the FESC in large numbers, a few individuals from Maine and New Hampshire were regular participants, and their concerns about the process of environmental assessment, research and model validation, as well as questions about mitigation measures and compensation, were shared by many New England coastal residents. At the June, 1982, meeting of the New England Governors' and Eastern Canadian Premiers' Conference, Joseph Brennan, Governor of the State of Maine, expressed concern about potential damage caused by the construction of a five-mile long tidal barrage in the upper Bay of Fundy. Subsequently, the Maine State Planning Office announced that funding for a preliminary environmental assessment study was being provided to researchers P.F. Larsen and J.A. Topinka, at the Bigelow Laboratory for Ocean Sciences in Boothbay Harbor, Maine.

The Larsen and Topinka report, released to the public in April, 1984, raised as many questions as it provided answers. A primary source of uncertainty cited by the authors was the fact that the range of *natural* variation in the Bay of Fundy is not adequately understood. Thus, prediction capabilities will remain limited until sufficient data are available to provide a baseline against which physical and biological alterations caused by a tidal power facility can be inferred. The "Greenberg model" developed by Canadian scientists[30] suggests that construction of a barrage in the Minas Basin site would result in a 15 cm (6 in.) elevation of mean high water levels and a 15 cm depression of mean low water levels. The Larsen and Topinka report discussed a wide range of potential impacts as a consequence of changes to the tidal regime, including: an estimated loss of 4200 acres of coastal property and territorial habitat; increased saltwater intrusion into groundwater aquifers; a doubling of the frequence of damaging storms; minor meteorological changes; increased estuarine flushing; increased vertical mixing in offshore regions, and shifting sedimentation patterns. Some of the changes were preceived to have potentially beneficial, as well as negative, effects; for example, increased flushing might assist in faster dispersal of spilled pollutants, but it might also cause pollutants to become embedded in bottom sediments in higher concentrations. The report concluded that additional research is necessary in *selected* areas including: phytoplankton and productivity; "fine tuning" the Greenberg model to identify responses of estuaries and bays to changes occurring throughout the system; quantification of projected impacts on migratory fish and bird species; and a detailed investigation of potential socio-economic impacts of tidal power on control communities through a case study approach. The report also emphasized the importance of considering alternative design features and operational procedures to determine whether or not changes at this level could mitigate susceptibility to damaging events.[31]

Concerns about adverse environmental impacts resulting from a large-scale tidal power project were again considered by the New England Governors and Eastern Canadian Premiers at their 1984 annual conference and at a pre-conference workshop on bilateral regional issues. Participants were involved in discussions about possible options for resolving potentially difficult conflicts arising from Fundy tidal power development, including: (i) a submission to the International Joint Commission (IJC); (ii) establishing a bilateral commission on marine resources and ocean boundaries; (iii) negotiation of a bilateral treaty with respect to the Fundy Tidal Project; (iv) creation of a Bay of Fundy/Gulf of Maine Scientific Committee under the auspices of the Governors and Premiers; and (v) amending existing U.S. and Canadian legislation to enable nationals of both countries to participate in respective environmental assessment and review processes.[32]

Concern about potentially adverse transboundary effects of Fundy tidal power extends beyond New England and has also become a federal issue. In July, 1983, a Committee of the U.S. Senate convened a public hearing to address questions about the "Effects of the Proposed Tidal Hydro-electric Project in the Bay of Fundy," and a bill was subsequently introduced (though never passed) that would authorize and fund the U.S. Army Corps of Engineers to undertake modeling studies in the Bay of Fundy.[33]

Annapolis Royal Demonstration Project

After 1977, changes in Canadian federal energy policies, as well as the escalation of production costs for alternative energy supplies, occurred at an unprecedented rate. The short-lived federal Progressive Conservative Government penned an agreement in January, 1980 with the Nova Scotia government and the Tidal Power Corporation (TPC) to contribute $25 million towards financing the estimated $46 million for a tidal power demonstration project on the Annapolis River.[34] The Tidal Power Corporation contracted a consulting firm, MARTEC Ltd., to assess and report on the environmental effects of constructing and operating a tidal power generating facility on the Annapolis River basin, with reference to assessment guidelines provided by the Nova Scotia Department of the Environment.[35] The MARTEC EIS document concluded that:

> The Annapolis Tidal Power Project can be constructed and operated with minimal impact on the regional environment, provided that measures to alleviate potential effects on marshland agriculture are implemented. The investigations have shown that, with proper planning and a few specialized studies to optimize water level management, the effects on the physical, biological, and human environments can be kept to a minimum and the residual impacts will be small. With proper planning, the region should enjoy significant short-term and some long-term economic benefits as a result of the development.[36]

The Report went on to recommend that a number of studies be carried out during the two-year construction phase to ensure that impacts associated with operating the project are "defined and controlled." These included water-level elevations, sediments and bathymetry, flood routing, agriculture, bank stability, and fish passage through the barrage and turbine operations. In addition, biological monitoring programs were recommended to fill in existing information gaps and to provide a baseline for post-operational monitoring[37] (see Table 2).

The MARTEC report was delivered to the TPC on February 1, 1980, but its findings were not simultaneously made public. Several days later, David Nantes, Chairman of the Tidal Power Corporation, announced that tenders would be called within the month for the Annapolis project, stating that tenders need not be held up because the MARTEC environmental impact study had not been circulated for public review. This announcement caused alarm, particularly in Annapolis County where residents voiced concern that impacts might indeed be more serious than anticipated. A local newspaper editorial summarized public sentiment:

> Since we must pay the environmental costs, then we insist that we should have a strong voice in the project now, not a few months down the road after a million or two has been committed. We urge concerned citizens to make this wish known to Mr. Kerr (Annapolis West MLA) and Mr. Buchanan (N.S. Premier). Public input must come first, it is our right as citizens of the area.[38]

In mid-February, 1980, members of the Nova Scotia Department of the Environment met with Tidal Power Corporation executives to discuss mutual concerns arising

Table 2

Risk Identification Matrix: Annapolis Tidal Power Project [1]

Identified Risks	Probability	Magnitude	Speed of Onset	Duration	Interactive Effects	Predictability	Population/ Resources at Risk
1. spoil disposal	High	High	Immediate	Short term		High	Benthic communities
2. basin scouring	Moderate	Unknown	Insidious	Continuous		Unknown	Benthic habitat
3. fish passage	Unknown	Unknown	Immediate	Episodic		Unknown	Striped bass, shad, salmon, gaspereau
4. water level management:							
a. bank erosion	Moderate	Unknown	Insidious	Episodic		Unknown	Agriculturalists
b. habitat disruption	Moderate	Low	Insidious	Continuous		Unknown	Waterfowl, fish, furbearers
c. increased groundwater levels	Low	Low	Insidious	Continuous	4e, 4f	Low	Agriculturalists
d. salt intrusion into subsoil	Moderate	Moderate	Insidious	Continuous		Unknown	Agriculturalists
e. impeded marshland drainage	High	Moderate	Immediate	Episodic	4f	High	Agriculturalists
f. flooding	Unknown	Unknown	Delayed	Episodic	4e	Unknown	Agriculturalists

Acknowledged data deficiency areas: (Section 5.3)

water level elevations
sediments and bathymetry
flood routing
agriculture: hydrogeology and groundwater quality; soil chemistry including nutrients, salinities, etc.; crop productivity and production limitations
bank stability
fisheries: fish pass operations; turbine operations
monitoring program
human environment

([1] MARTEC Ltd., Annapolis Tidal Power Project Environmental Impact Assessment, 1980.)

from the MARTEC report. Specifically, the project proponent was instructed in the environmental assessment guidelines to "define beneficial and adverse environmental impacts of all feasible alternatives with specific reference to the existing environment in the project area(s) including: agriculture, fisheries, water, mineral and construction materials, recreation and forestry."[39] The MARTEC report described existing fisheries in terms of stock status, but the impact statement failed to discuss possible impacts that construction or generating operations might have for the area's commercial and sports fisheries. The result of the informal consultation meeting was that the Minister of the Environment gave conditional approval to the project with the proviso that the Tidal Power Corporation would agree to undertake additional studies. In addition, an Intergovernmental Environmental Advisory Committee was established in June, 1980 to monitor and assess work conducted on behalf of, and by, the Tidal Power Corporation.[40]

As the Annapolis project proceeded, several local interest groups began questioning developers about potential impacts of construction and operation. These groups included the Annapolis County Federation of Agriculture, the Concerned Homeowners and Citizens of the Annapolis River Tidal Areas, and the Annapolis District Planning Commission. Whereas these groups were peripheral to the MARTEC and TPC assessment process, they were instrumental in bringing public attention to issues which project proponents had preferred to ignore. However, by 1980, the prevailing perception among Annapolis County residents was that it was too late to stop the project, but some form of mitigation, remedy or compensation might be worked out: "(i)t's too late to head off any adverse effects from the power plant; the tidal power project is proceeding regardless. What (we) want now is to be informed in advance of what may happen so (we) can work out with the province how to deal with it."[41]

After lengthy construction delays caused in part by labor strikes at the site, in May, 1984, the Straflo turbines at Annapolis were tested for a full tidal cycle, and the first tidal-generated electricity was produced in North America. Post-testing engineering adjustments were completed during the summer and the Annapolis Royal demonstration project was officially opened in August, 1984. The opening, however, was not greeted universally with enthusiasm, and while ribbons were being cut by government officials, local farmers protested by staging a tractor parade and rally across the Annapolis causeway. Although the Annapolis project is now fully operational, there are still no firm commitments to compensate for damages or formal procedures to appeal for compensatory payment; thus, local homeowners, farmers, or fishermen who suffer injury or loss as a consequence of tidal power generation are left without a streamlined legal course of recovery.

Discussion

In summary, the issues raised by the Annapolis case study and the evolving Fundy study indicate that technology assessment systems do not always function in ways that promote comprehensive, unbiased, and critical evaluations. Concerns about Canadian federal and provincial assessment practices raised by the studies include the following points:

1. Assessment systems are selective. Assessments conducted by (or on behalf of) project proponents tend to reflect the proponents' perception of what information is necessary and relevant to decisionmakers to the exclusion of other perceptions.

2. Assessments can be accelerated by political momentum. Political interest and authority influence the pace and procedures of the assessment process.

3. Information can be controlled and debate limited. Although proponents' interest may stimulate research in previously unstudied areas, proponents are privy to study findings prior to public release. In addition, circulation of results and provision of forums for debate become private decisions rather than a matter of public responsibility.

Doern (1981) argues that processes for obtaining scientific and technological data and advice are inadequate within the Canadian federal government, suggesting that: "one is forced to conclude that scientific and technological controversies are, on balance, viewed by decisionmakers as peripheral to the formal decision process in comparison with competing political, economic, legal and other elements. This is partly due to the problem of definition and perception."[42] The Fundy tidal power case study demonstrates that Doern's analysis is also applicable at the provincial level where "debate" about tidal power development has focused on economic feasibility. The primary obstacles to development, as perceived by decisionmakers, are construction costs and uncertain markets. Environmental risks are *not* perceived by proponents as posing a threat to the initial decision to develop a tidal power facility. The repeated failure to undertake comprehensive environmental impact studies during the course of assessment and reassessment investigations suggests that regional decisionmakers perceive that: (1) environmental risks are of second-level significance in terms of project approval, and (2) engineering adjustments are capable of minimizing negative consequences incurred by construction or during the project's operational life. Most significant, however, is the fact that the concept of risk was totally absent from the agendas of several tidal power inquiries. Since risk assessment is problem-oriented, it implies weighing costs and benefits before making decisions. As the Annapolis and evolving Fundy case studies illustrate, the perception that environmental risks should be equally weighted with financial and technological risks is not prevalent at levels where decisions are made -- in the political realm.

These case studies also illustrate how potentially at-risk resources can be identified, yet effects can be ignored at the decision stage. The failure to incorporate compensation, mitigation, and remedial works issues into the decision process -- at least in the case of the Annapolis project -- does not generate public confidence in regulatory environmental review or political decision processes. Because many of the impacts of tidal power remain unknown, the Tidal Power Corporation has preferred to adopt a "wait and see" attitude. For example, the Executive Vice-President of the TPC stated at an April, 1983, public meeting in Annapolis that: "If the Tidal Power Corporation is responsible for any property damage done as a result of accelerated erosion, we will pay landowners. But we will not compensate for natural erosion."[43]

Responsibility for making such evaluations remains an open question, and appeal and monitoring procedures have not been finalized or made public. Cottage owners were told their problems would be addressed on a one-to-one basis, and fishermen were left with as many questions as they raised about the impacts of the turbines on their vessels: "We will have to wait and find out what the water conditions will be. If you (fishermen) can moor there, then fine. If not, then we will have a problem and we can deal with it then."[44]

The Annapolis project is, however, an important component in the long-term assessment and planning process for harnessing Fundy tidal power. As concern intensifies in the United States about possible environmental, social, and economic effects of a large-scale tidal power generating facility, Annapolis provides an opportunity to evaluate local-level impacts and to refine designs for remedial works. The delays in making the decision to proceed with a large-scale tidal power plant, caused initially by economic factors, has permitted interests outside the political arena to become more actively involved in the assessment process. It is anticipated that future environmental assessment activity will be more rigorous and will benefit from the wealth of data accumulated by the Fundy Environmental Studies Committee, as well as by scientists and research institutions across the Canada - U.S. border.[45]

Final decisions to proceed with tidal power development will be scrutinized by a wider and better informed audience as a result of the protracted evaluation process. Most important may be the fact that the politics of tidal power development have changed so that approval will be contingent on reaching a regional and bi-national consensus concerning the acceptability of risks associated with development, and on shared perceptions that the benefits of tidal power generation outweigh the costs. For example, George Baker, Executive Vice President of the Tidal Power Corporation, told an audience in September, 1985, that, "In the long run, however, a Minas Basin project may not be 'developable' from a political rather than a commercial, point of view."[46] However, the decision to develop a large-scale tidal power generating facility will not be made in isolation from other factors, including cost-benefit and risk analysis of alternative Atlantic energy options such as hydro, coal, natural gas, oil, and nuclear/uranium[47] (see Table 3). Finally, the creation of mutually acceptable institutional or legal mechanisms to resolve project generated disputes is now almost certainly a pre-condition to project construction.

Table 3
Some Comparative Aspects of Atlantic Canada's Energy Options

	Tidal	Hydro	Coal	Natural Gas	Oil	Nuclear/Uranium
Location	N.B./N.S.	Labrador	N.S.	N.S. (Sable) Labrador?	Nfld. (Hibernia)	N.B./N.S.
Type	Electricity (Peaking Power) - Intermittent	Electricity Peaking; Some Base Load	Electricity/Gas (Base Load and Peaking Power)	Heating/ Convertible to Electricity	Heating, Petro-chemical Supply, Aviation Fuel/ Convertible to Electricity	Electricity (Base Load)
Markets	Local, Eastern U.S.	Eastern U.S., local	Local	Local, Eastern U.S.	Local, Eastern U.S.	Eastern U.S., Local
Quantity	4,000 MW (B9)		$6x10^9$ tonnes			2,400 MW (assuming 4 units at Pt. Lepreau)
Costs	\$6 billion (B9) \$1.8 billion (A8) (Excl. transmission lines 1.2 billion	\$4.3 billion (Excl. transmission costs of 1.6 billion)	1.5 billion	uncertain	uncertain	\$5 billion?
Employment		10,000 man-years for construction place.	7,000 coal-related jobs			
Interference with Other Resources	Agriculture and Forestry (Transmission Lines)	Agriculture and Forestry (Transmission Lines)		Fisheries (Well head and Production Platforms)	Fisheries (Well head and Production Platforms)	Argiculture and Forestry (Transmission Lines)
Transport	Transmission Lines	Transmission Lines	Road, Rail, Sea (Transmission Lines for Conversion to Electricity)	Undersea & Land Pipelines	Tankers? Undersea & Land Pipelines	Transmission Lines

	Tidal	Hydro	Coal	Natural Gas	Oil	Nuclear/Uranium
Air Pollution	None	None	Acidic Precipitation (Acid Rain)	None	None	None
Water Pollution	None	None (Construction Damage	-Metal & Acid Leaching -Acid Precipitation -Thermal Pollution	In Accidents	In Accidental Spills	-Radiation Release in Accidents -Thermal Pollution
Environmental Changes	-Modification of Tidal Ranges in Bay of Fundy/ Gulf of Maine -Loss of Inter-tidal Zone -Stratification of Headpond -Reduced Growing Season	-Loss of River and Forest Habitat -Fish Passage -Microclimate Effects	-Reduction of Soil Productivity -Destruction of Lake and Stream Fisheries	-Accidental Destruction of Fish, Plankton -Damage to Fish Spawning Grounds	-Accidental Destruction of Fish, Plankton Benthos -Damage to Fish Spawning Grounds	-Thermal Pollution -Accidental Destruction of Fish & Plankton
Legal Aspects	-International (Canada/US): -Tidal Changes -Migratory Fish & Birds -National Export	-National: - Export Licenses - Interprovincial provincial Transportation	-International: -Acid Rain (Canada/ Greenland) -National: Acid Rain (NS/NFLD)	-International: -Georges Bank (Canada/US) -Fisheries -National: Export Licenses	-International: -Georges Bank (Canada/US) -Grand Banks Fisheries -National: Export Licenses	-National: Export Licenses
Other	-Export Canadian Technology -Renewable -Life 75 + years	-Remoteness From From Markets -Renewable -Life: 25 + years	-Waste Disposal -Non-Renewable -Life 200+ years	-Sensitive to World Prices -Boost to Petro-Chemical -Life 20-25 yrs	-Sensitive to World Prices -Boost to Petro-Chemical -Life: 20 + yrs	-Export Canadian Technology -Insensitive to World Prices -Life: 50 years?

Notes

1. Dorothy Nelkin and Michael Pollak, "Consensus and Conflict Resolution: The Politics of Assessing Risk," in *Technological Risk*, Meinoff Dierkes, Sam Edwards, Rob Coppock (eds.), Cambridge: Delgeschlager, Gunn and Hain, Publishers, Inc. (1980), pp. 65-75, at p. 65.

2. *Ibid.*, pp. 65-66. See also, G. Bruce Doern, *The Peripheral Nature of Scientific and Technological Controversy in Federal Policy Formation*. Science Council of Canada. Background Study 46 (1981).

3. *See*, for example, Liora Satter and Debra Slace, *Public Inquiries in Canada*. Science Council of Canada. Background Study No. 47 (1981); L. Graham Smith, "Alternative Mechanisms for Public Participation in Environmental Policy-Making," *Environments* V. 14, No. 3 (1982) pp. 21-34; and John E. Carroll, *Environmental Diplomacy: An Examination and a Prospective of Canadian-U.S. Transboundary Environmental Relations* (1983).

4. *See*, for example, *Northern Development and Technology Assessment Systems: A Study of Petroleum Development Programs in the Mackenzie Delta-Beaufort Sea Region and Arctic Islands*, George R. Francis and Sally C. Lerner. Science Council of Canada. Background Study No. 34 (1976).

5. Dexter P. Cooper Inc. promoted plans for an international two-basin scheme using the tides of Passamaquoddy and Cobscook Bay; in 1935, the U.S. began construction of a facility on Cobscook Bay but suspended operations when project funding ceased. The governments of Canada and New Brunswick investigated a site at the mouths of the Petitecodiac and Memramcook Rivers in 1944, and the International Joint Commission investigated the costs of developing tidal power in Passamaquoddy Bay in 1956. The Atlantic Development Board sponsored surveys in the Shepody and Cumberland estuaries and Minas Basin from 1963 to 1966. *See* Atlantic Tidal Power Programming Board, *Report*, 1969: 9-10.

6. *See* Atlantic Tidal Power Programming Board, *Report*, 1969: 28.

7. *Ibid.*, p. 159.

8. *Ibid.*, p. 170.

9. "Fear Quebec Will Turn Tide Against Fundy Development," (Halifax) *Mail-Star*, April 29, 1971; "Hydro-Quebec Could Undercut Tidal Power," (Halifax) *Chronicle Herald*, January 31, 1984.

10. "Fundy Power: 'Further Studies Necessary,' " (Halifax) *Chronicle Herald*, January 31, 1984.

11. "Ottawa Agrees to Review of Tidal Study," (Halifax) *Mail-Star*, April 1, 1971.

12. *Ibid.*

13. Bay of Fundy Tidal Review Board, Preliminary Reassessment of the Feasibility of Tidal Power in the Bay of Fundy, *Report*, 1974: 11.

14. Bay of Fundy Tidal Review Board, *Reassessment of Fundy Tidal Power* (Report), 1977.

15. *Ibid.*, p. 385 (emphasis added).

16. *Ibid.*, p. 387.

17. *Ibid.*, p. 390.

18. *Ibid.*, pp. 390-391.

19. *Ibid.*, p. 391.

20. *Ibid.*, pp. 387-389.

21. *Ibid.*, p. 393.

22. *Ibid.*, p. 404.

23. *See* Tidal Power Corporation. *Fundy Tidal Power: Update '82* (1982).

24. "More Funding for Tidal Power Studies," (Halifax) *Chronicle Herald*, May 24, 1984.

25. "Alcan Studies Investor Interest in Fundy Tidal Power Project," (Halifax) *Chronicle Herald*, July 1, 1985.

26. Donald C. Gordon, Jr. and Alan R. Longhurst, "The Environmental Aspects of a Tidal Power Project in the Upper Reaches of the Bay of Fundy," *Marine Pollution Bulletin*, v. 10 (1979), pp. 38-45.

27. "Fundy Tidal Power and the Environment," *Proceedings* of a Workshop on the Environmental Implications of Fundy Tidal Power, Wolfville, N.S. (November 4-5, 1976), p. iii.

28. Information provided through interviews conducted by Peter N. Duinker with Dr. D.C. Gordon, Jr. (FESC Chairperson, 1977-82); Dr. G.R. Daborn (current chairperson, FESC), and Dr. F.J. Simpson (APCS chairperson), in the summer of 1983. A valuable compendium of information is available in *Update of the Marine Environmental Consequences of Tidal Power Development in the Upper Reaches of the Bay of Fundy*, Donald C. Gordon and Mike C. Dadswell (Eds.) Canadian Technical Report of Fisheries and Aquatic Sciences - No. 1256 (1984).

29. *See Regional Scientific Needs in Fundy Tidal Power Development*, Report of a workshop held at the University of New Hampshire (November 20-21, 1983). Not all scientists share this view of modeling predictions, see *Fundy Tidal Power Development: Preliminary Evaluation of its Environmental Consequences to Maine*, P.F. Larsen and J.A. Topinka (eds.). A Report to the Maine State Planning Office by the Bigelow Laboratory for Ocean Sciences (1984).

30. *See* G.F.D. Duff, "Numerical Modeling of Tides in the Bay of Fundy," *Tidal Power and Estuary Management*, R.T. Severn, D. Dineley, L.E. Hawker (Eds.) Bristol: Scientechnica (1979), pp. 93-98.

31. *See Fundy Tidal Power Development: Preliminary Evaluation of its Environmental Consequences to Maine*, *supra* note 29.

32. *See Background Paper on the Proposed Bay of Fundy Tidal Power Project*, prepared for the New England Governors' Conference for First Annual Bilateral Symposium on New England Eastern Canadian Affairs (Providence, Rhode Island, May 24-25, 1984).

33. U.S. Senate 98th Congress, 1st Session. "The Effect of the Proposed Tidal Hydro-Electric Project in the Bay of Fundy," *Proceedings* of a Hearing before the Committee on Environment and Public Works (Augusta, Maine, July 25, 1983). The U.S. Army Corps of Engineers, New England Division, sponsored a public meeting to discuss its activities (October 30-31 and November 1, 1984 at Rockport, Maine).

34. "Tidal Power Experiment Agreement Today," *Chronicle-Herald*, January 29, 1980. By 1982, the cost of the project was approaching the $52 million mark.

35. "Guidelines and Information Requirements for an Environmental Impact Assessment of the Annapolis Royal Tidal Power Station, Annapolis Royal, N.S.," Nova Scotia Dept. of the Environment, Environmental Assessment Division (August, 1979), File No. 1750-A3.

36. MARTEC Ltd., *Annapolis Tidal Power Project: Environmental Impact Assessment*, 1980, 5:11.

37. *Ibid.*, p. 5-11 - 5:16.

38. "Public Input First," *Mirror*, February 13, 1980.

39. *See* "Guidelines and Information Requirements, *op. cit.*

40. File Documents, Nova Scotia Department of the Environment, 1750-A3 and 1570-3 (April 11, 1980), and 1570-2 and 1750-A3 (June 11, 1980).

41. "Voice Concern over Impact," *Mirror*, May 7, 1980.

42. G. Bruce Doern. *The Peripheral Nature of Scientific and Technological Controversy in Federal Policy Formation*. Science Council of Canada. Background Study No. 46 (July, 1981), p. 88.

43. "Fishermen Get Few Answers to Concerns About Fundy Power," *Chronicle-Herald*, April 15, 1983.

44. *Ibid.*

45. "Greater U.S. Input Needed on Fundy Effects -- Daborn," (Halifax) *Chronicle-Herald*, June 1, 1984.

46. "Environmental Politics May Kill Tidal Project," (Halifax) *Chronicle-Herald*, September 11, 1985.

47. G.R. Daborn and M.W. Couley (Eds.), *Energy Options for Atlantic Canada*. Halifax: Formac Publishing Co.,Ltd in association with Acadia University Institute (1983).

Chapter 2

Scientific Issues in the Assessment of the Transboundary Effects of Fundy Tidal Power

Judith Spiller

Introduction

Over the past fifty years, several proposals have been advanced to harness the exceptionally high tides of the Bay of Fundy region for the generation of electricity.[1] Each proposal, however, has been rejected on the grounds that it was uneconomical. Changes in the world energy supply, culminating with the Arab oil embargo in the 1970s, have however stimulated serious reconsideration of tidal power.[2] That reconsideration has included examination of the environmental consequences of a large-scale tidal energy facility constructed in the Bay of Fundy in Nova Scotia.

When considering a new technology that impinges on an ecosystem where the background level of change is poorly understood, a significant problem exists in assessing the potential effects. Effects close to the project site are relatively easier to evaluate because cause and effect relationships may be established and analogs from other areas employed. Some evidence on the potential near-field effects of tidal power, for example, comes from sites where causeways have been constructed.[3] Similarly, information on the effects of turbine operation and the efficacy of fishways may be employed to estimate fish population changes.

More difficult to evaluate, though, are the longer-term and more distant changes in ecosystem structure. Computer simulation of the Bay of Fundy-Gulf of Maine region suggests that tidal ranges and currents will be affected throughout the Bay and Gulf.[4] Numerous consequences (Table 1), most prominent being extensive coastal flooding and alterations in fisheries productivity, have been predicted to result from the shift in tidal response indicated by the modeling. Inability to predict ecosystem behavior makes assigning probabilities to those consequences an imprecise task at best. This uncertainty has fueled local concern about the environmental implications of Fundy tidal power, particularly along the Maine coast.

Analysis of potential effects must focus on two questions. First, do other system perturbations such as meteorological and astronomical events result in changes similar to those speculated for Fundy tidal power, and second, does the natural variability of the system (the Gulf and the Bay) exceed the magnitude of changes suggested as a consequence of the project? If the proposed changes in tidal amplitude and currents are only small increments over what is currently observed, then they will have negligible transboundary effects. Similarly, if the far-field effects of the project fall within the range of existing variation, they will be system-wide and thus undetectable. If, however, these alterations are magnified by phenomena like storms and cyclical tidal events, then their effect on the Gulf and Bay environments may exceed observed variation. In this case, changes must be assessed as to their potential benefit or harm, but their relation to the system variation must be established first.

Table 1: Potential Consequences of Fundy Tidal Power Development[5]

A. Change in Tidal Range (~10%)

1. increase in intertidal area
2. expanded intertidal zonation (~10%)
3. change in intertidal community structure
4. intertidal sediment translocation
5. supratidal sediment translocation
6. increase in salt marsh area (~10%)
7. increase in size of estuarial tidal prism (~10%+)
8. increase in area for ice formation (5–10%)
9. change in groundwater hydrology
10. shoreline submergence
11. shoreline erosion (beaches)
12. tidal inlet stability effects
13. increase in flooding, increase in frequence of storm surges
14. loss of subtidal habitat

B. Change in Tidal Current (~10%)

1. change in mixing
2. increase in nutrient supply to surface waters
3. increase in vertical transport
4. change in surface water temperature
5. change in meteorology from change in temperature
6. life history change from temperature change
7. altered fish migrations from temperature change
8. increase in abundance of fouling organisms
9. increase in flushing, larval dispersal
10. restructure plant/animal communities
11. increase in mixing, transport and abundance of red tide toxins
12. increase in transport/dispersal of pollutants
13. inhibit ice formation due to increased mixing
14. increase in ice transport

C. Changes in Productivity

1. increase in primary productivity
2. increase in seaweed production (~10%)
3. increase in intertidal saltmarsh productivity (~10%)
4. biological readjustment

D. Uncertain Impacts

1. fisheries (shrimp, lobster, crab, scallop)
2. clam, worm, mussel harvesting
3. fin fishing (dragging in increased currents)
4. anadromous fishing
5. intertidal/subtidal seaweed collection
6. aquaculture

Evaluating the potential effects of Fundy tidal power involves three steps: definition of the extent of possible change, examination of the variability of the forces and linkages that structure the area's ecosystem, and delineation of the categories of expected effects. This chapter defines the geographic extent of the effects attributed to the proposed project and the variability of the forces and linkages which structure the Bay and the Gulf. It presents an analysis of the basis for predicting changes in the Bay and Gulf tidal regimes and examines the categories of transboundary effects attributed to the proposed project. The environmental effects associated with other large-scale tidal power projects are also considered. The chapter concludes with a discussion of priority areas for future research.

Extent of Effects

Boundary of the Ecosystem

Analysis of the scientific research relevant to the issue of the transboundary effects of Fundy tidal power begins with defining the extent of effects. Computer simulation of the region's tidal regime, with and without tidal barriers, indicates alterations in tidal amplitudes and phase changes from the proposed barrier sites, the preferred B9 and the alternate A8 site, in the upper Fundy to Georges Bank (See Figure 2 in chapter 1, page 5). Physical boundaries of the Bay of Fundy-Gulf of Maine basin largely coincide with the area of effects predicted in the model. The boundary between the Gulf and Georges Bank, including the Great South Channel and the Northeast Channel, forms the southeastern limit, and the coastlines of New Brunswick, Maine, New Hampshire, Massachusetts and Nova Scotia enclose the rest of the area (Figure 3).

Linkage and Variability

A fundamental question raised in assessing transboundary effects is the degree to which the Bay and the Gulf act as an integrated system. The natural oscillatory period or the functional mode (the time it takes a wave to travel the length of the Bay and the Gulf) of both water bodies must be considered jointly to derive the model's predictions that simulate the observed tides. Tides thus join the Bay and the Gulf as does mean circulation (residual nontidal water movement). Mean circulation forms a counterclockwise gyre circumscribing the Gulf of Maine and driven primarily by water from the Northeast Channel, secondarily by the Great South Channel, and occasionally by excursions of the Georges Bank water mass.[6] Some water from this gyre diverts into the Bay of Fundy, contributing to the counterclockwise mean circulation there.[7] Before rejoining the Gulf waters, this circulation moves water up the Bay along its western edge and out along its eastern side. Upwelling occurs where the Gulf and Georges Bank water masses meet, around the topographic highs of the Gulf's fishing banks, and along the southwest Nova Scotia coast.

The breakdown and reconstruction of the hydrographic structures depends on the amount and timing of the releases of freshwater from the St. Lawrence River,[8] the strength of the Gulf gyre, and the behavior of water on Georges Bank. Residual circulation changes seasonally (Figures 4 and 5).

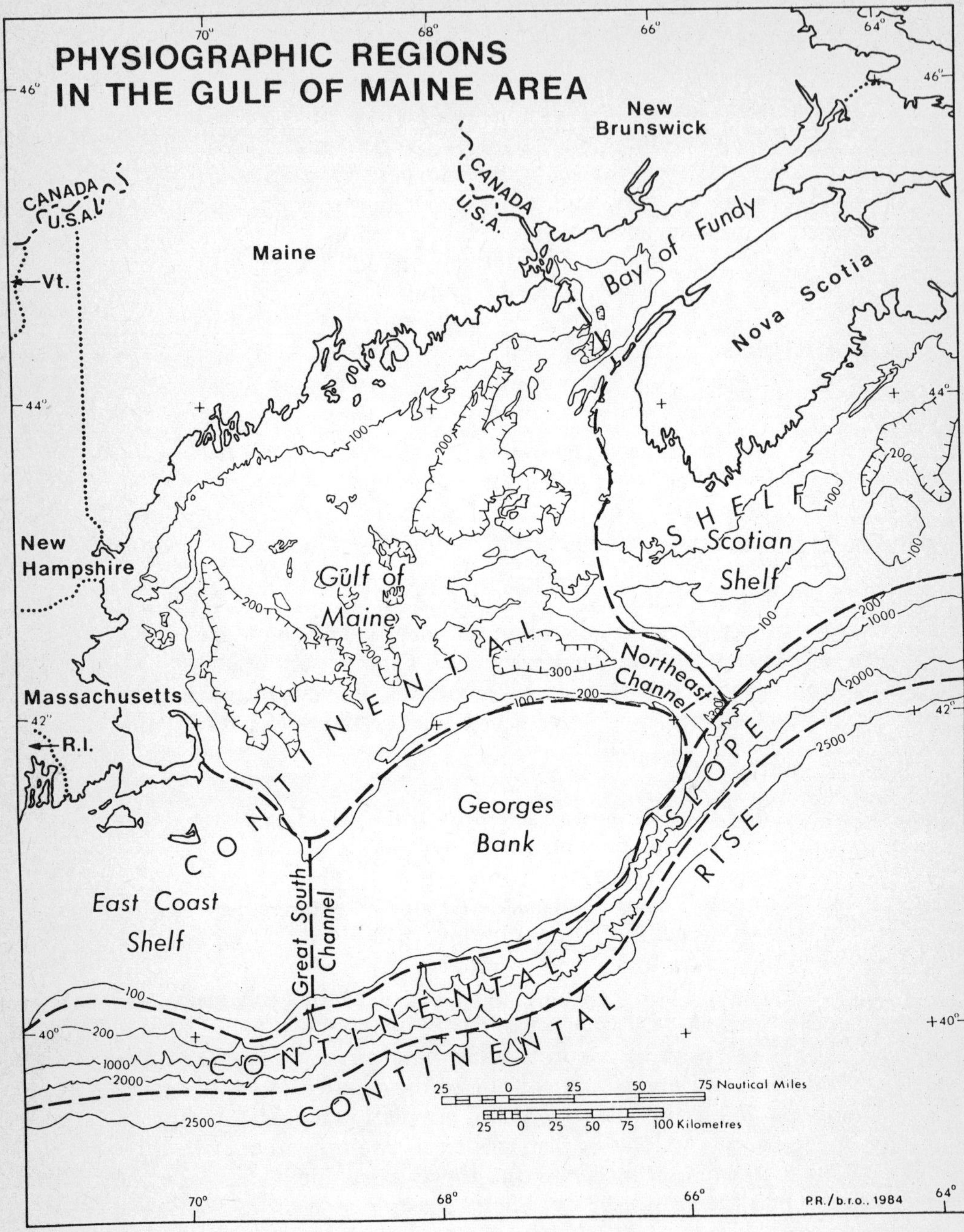

Figure 3. Physiographic Regions in the Gulf of Maine.

Map prepared by P.J. Ricketts, Saint Mary's University and J. Hutchings, Dalhousie University. Published with permission.

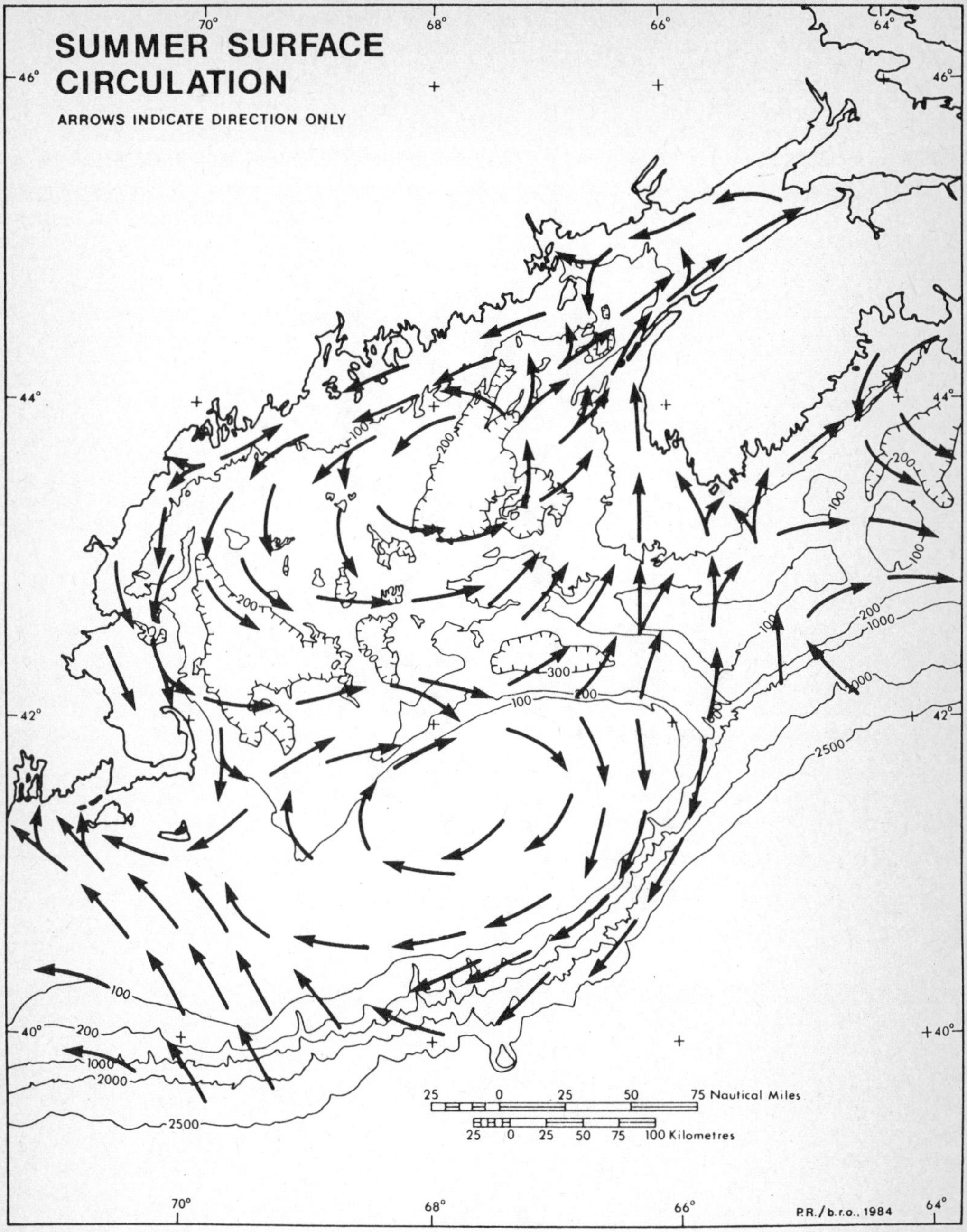

Figure 4. Summer Surface Circulation in the Gulf of Maine Region.

From Hutchings, J. and P.J. Ricketts, "Mapping the Environmental Resources of the Bay of Fundy, Gulf of Maine and Georges Bank (FMG) Marine Region" in D. Day and P.J. Ricketts (eds.), *Studies in Marine and Coastal Geography No. 3 (1984), p. 32 (Original Source:* Bumpus, D.F., "A Description of the Circulation on the Continental Shelf of the Eastern United States" in *Progress in Oceanography* No. 6 (1974), pp. 111-157).

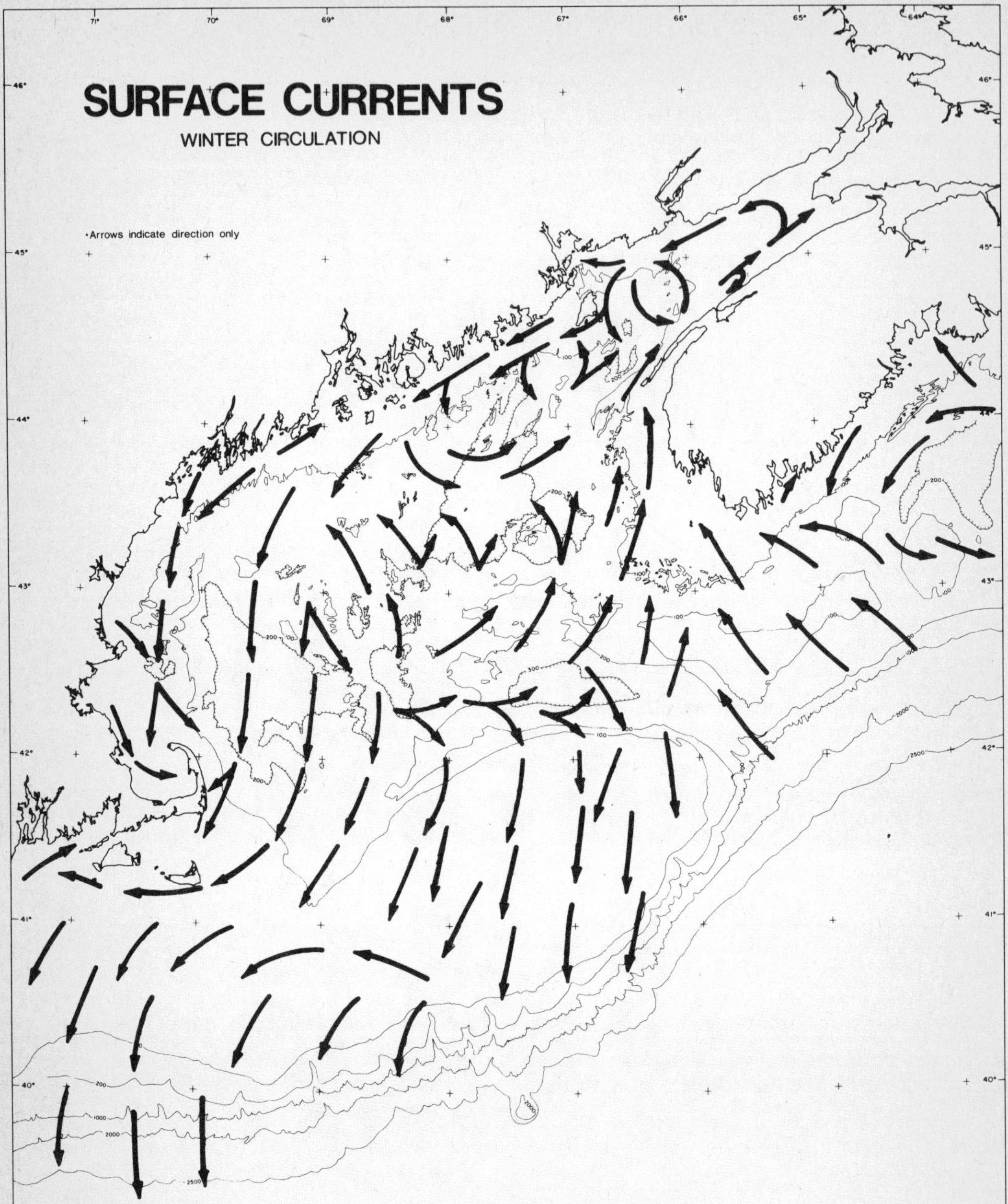

Figure 5. Winter Surface Currents in the Gulf of Maine Region.

Map prepared by P.J. Ricketts, Saint Mary's University and J. Hutchings, Dalhousie University, based on Bumpus, D.F., "A Description of the Circulation on the Continental Shelf of the Eastern United States" in *Progress in Oceanography* No. 6 (1974), pp. 111-157.

Tidal action varies with numerous astronomical cycles and meteorological forces. Bottom topography, coastal configuration, wind, freshwater inputs, and water mass characteristics all impinge on mean circulation and tidal movements. These changes in circulation also influence water temperature.

Temperature and salinity interact to create density structures in the water column. Solar radiation plays a significant role, particularly in shallower waters. Spring and summer warming of the upper layers makes temperature differences at various depths become more pronounced and creates vertical stratification in the water column.[9] During the colder months, the water becomes more homogeneous. Salinity values also show seasonal variation, reaching their maximum during the late fall. In addition, freshwater inputs and more saline waters, entering from the continental slope, influence these values.[10]

Physical forces in the Gulf and Bay strongly influence the region's biological components. For example, the interface between the Gulf and the Bay around Grand Manan Island forms an area of complex mixing, controlled by residual and tidal currents and the local topography. Here, contact between the Fundy and Gulf waters forms a small counterclockwise gyre. When the gyre is strong, it cuts off some water exchange with the Bay, maintaining organisms in that area. At other times when it weakens, it allows for dispersion of the biota. Vertical stratification and upwelling permit organisms to exploit resources at differing levels in the water column and at various times of the year. Smaller organisms, like the zooplanktonic euphausiids, concentrate in the areas of contact, remaining at specific depths and areas.[11] Other organisms such as herring congregate in the area seasonally and then disperse.[12] Finally, larger organisms such as shad[13] and some whales[14] migrate through the region.

In summary, the biological structure of the Bay of Fundy and the Gulf of Maine is largely controlled by the physical system of gyres and upwelling (vertical mixing). The location and the persistence of these features is, on the average, constant. Although the Gulf of Maine gyre usually weakens in the winter, the degree of weakening is unpredictable. This unpredictability may also be seen in the strength and location of the gyres around Grand Manan Island and in the extent of upwelling along the southwest Nova Scotian coast. As a consequence of both the variability of the region's physical parameters and the inherent variability of the biota (e.g., its reproductive condition), the time of species appearance and the year class strength fluctuate. Thus, although events in the physical system and in organisms' life histories are predictable in a general sense, more specific hypotheses concerning events (e.g., the time of spawning or year class of strength) are easily invalidated by other observations.

Computer Simulation of the Gulf of Maine-Bay of Fundy Tidal Regime

Validation with field measurements and other modeling[15] indicates that computer simulation[16] of the tidal regime for the Gulf of Maine and the Bay of Fundy region closely represents its tidal properties. The model prepared by Dr. David Greenberg of the Bedford Institute of Oceanography (a two dimensional, time-dependent, finite difference model) predicts that a tidal barrier placed at the preferred B9 site at the mouth of Cobequid Bay will result in tidal amplitude increases ranging from 9 centimeters to 15 centimeters along the coasts of Maine, Massachusetts, New Brunswick, and Nova Scotia. The nature and extent of the effects of these increases on coastal habitats and marine resources depends on the behavior of a number of physical forces. Analysis of how these forces are resolved in Greenberg's simulation provides insight into their potential for predicting possible effects. The examination of the Greenberg model here begins with a brief description of the physical system and the forces controlling the region's tides, continues with an examination of the parameters used in the modeling, analyzes the regime produced by the model, and finally, explores the effects of barriers as predicted by the model.

The Gulf of Maine-Bay of Fundy System

Tides occur because of the contemporaneous effects of the gravitational forces of the sun and the moon and the rotation of the moon about the earth and the earth about the sun. The moon exerts the principal tide-producing effect. Its gravitational force acts to draw water on the near side of the earth toward the moon, causing a rise in tides. These parallel forces also act on the side of the earth opposite to the moon, drawing the earth away from the water mass and raising tides. Areas perpendicular to the moon's parallel lines of force, and so opposing those experiencing high tides, will be subject to low tides.

At any given time, different parts of the earth are at differing distances from the moon; therefore, the attraction of gravity varies. Further, the moon moves around the earth following a 27.3 day period, and the earth also rotates, resulting, with respect to the moon, in a 24.8 hour lunar day. Over this lunar day or period, the earth will experience semi-diurnal tides - two high and two low tides. The moon's forces are greatest where they are nearly parallel to the earth, creating the tidal bulges of water. The movement of the earth and the moon, however, changes the tidal bulges' location. Friction between the ocean and the ocean bottom slows the bulge down as do continental land masses.[17] Thus, for example, high tides average five to six hours behind the time when the moon is directly overhead.

Other forces also affect tides in open oceans. The sun sets up a gravitational effect approximately half of that of the moon.[18] Further, the sun and moon do not rotate synchronously. Their planes of rotation are not in an equatorial plane.[19] The sun moves north and south on an annual cycle. The moon moves similarly but on a monthly cycle. Each of these exerts a force on the tides. At least eleven other factors must be considered in predicting tidal effects in open waters, and as many as sixty-five may be necessary to characterize tides in estuaries.[20] The principal tide-producing effect, however, is the moon.

Tidal prediction is based on knowledge of the periodicities of the major tide-forcing elements. This information derives from years of observation of tidal behavior at various geographic locations for periods of time. Based on these data, tidal height curves are determined and are then extrapolated into the future.[21] These curves are the sums of a number of harmonics (waves which express the periodicities and amplitudes of the major tide-forcing elements). Their projection involves a degree of error which increases as predictions go further into the future. These predictions for the open ocean come within about 10% of accuracy.[22]

Shape of the ocean basin and meteorological conditions also influence tidal prediction. Tide-producing forces may respond more readily in some basins or with the passage of a storm than they will in other areas or under different weather conditions. Basin configuration effects may be accounted for after observation. Meteorological tides which include atmospheric pressure and set-up (slope of the water surface due to wind), however, cannot be predicted.[23] Except for large storms like hurricanes, atmospheric disturbances may cause changes of ± 10 centimeters (cm) and ± 10 minutes in tidal amplitudes.[24]

The Gulf of Maine is essentially a rectangular basin, bounded by Maine and New Brunswick on the west, the continental shelf on the east, Cape Cod on the southern edge, and the Bay of Fundy and Nova Scotia in the north. Seaward portions encompass Georges Bank, where depths are as shallow as a few meters, and Browns Bank/Scotian Shelf. Water enters this basin from two sources. One, the Fundian or Northeast Channel joins the Gulf with the open ocean and separates Georges Bank from Browns Bank. It deepens to 360 meters in places and represents the only deep entrance to the Gulf. The other, the Great South Channel, divides the western edge of Georges Bank from Cape Cod and reaches depths of 140 meters. The main body of the Bay of Fundy which extends beyond the northeastern section of the Gulf of Maine is about 170 kilometers (km) long.[25] Its width varies from about 48 km at Cape Chignecto to 120 km at the mouth, opening out into the Gulf of Maine. The depths along the center line of the Bay shallow gradually from 180 meters near the Gulf of Maine to 40 meters at the head where the Bay of Fundy divides into Chignecto Bay and Minas Basin. In contrast, the Gulf of Maine reaches depths of 350 meters.

The tides in the region are driven by those on the open ocean boundary along Georges Bank. Astronomical forces act primarily on large bodies of water - the water in major ocean basins rather than in the smaller adjacent bays and gulfs.[26] The tidal response of the open ocean forces the tides in shallower, neighboring bodies of water. Thus, water is forced from the North Atlantic into the region through the Northeast Channel where water depths change from over 2,000 meters beyond shelf to 200 meters at the inner edge of the Channel. This water feeds the counterclockwise gyre circumscribing the Gulf. Another smaller infusion of water enters through the Great South Channel.

The distinct tidal response of the Bay-Gulf system results from the nature of its boundary with the open ocean and the resonant condition created by the near coincidence of its natural oscillatory period and the lunar tidal period. Near resonant conditions in the Bay force tides varying from 2.4 meters near the mouth of the Bay of Fundy to 16.3 meters at the head.[27] Garrett[28] first hypothesized that the natural oscillatory period approached the driving period of the principal lunar tide designated as M_2. It takes 12.42 hours for the lunar tide to complete a cycle from one high tide to the next. Recent investigations[29] place the natural oscillatory period of the Bay and Gulf at about 12.5 - 13.00 hours. The similarity between these periods provides the conditions for resonance, and together they reproduce the observed tidal heights. The conditions approaching resonance and impelling the observed tidal heights require the inclusion of the Gulf of Maine in addition to the Bay in these calculations.

The relation between tidal forcing, the natural oscillatory period, and the lunar period is critical to predictions concerning the effects of tidal power. Barriers which shorten the axis of the system would reduce the natural oscillatory period. If the natural period were less than the lunar tidal period, then its reduction after the emplacement of barriers would move the natural period further from the M_2 period, and resonant effects would be diminished. If, on the other hand, the natural period were greater than the lunar period, then the emplacement of barriers would move the two periods closer together, creating greater resonance and increasing tidal amplitudes.

Current modeling efforts have grown from a number of studies. Initial research (Table 2) attributed the high tidal range of the region to resonance of the Bay alone with the principal lunar tide, M_2 mode. Proudman, using a constant depth representation, calculated a tidal period of 11 hours and 5 minutes for the Bay of Fundy.[31] A more detailed study by Rao contradicted previous notions of the Bay's resonance with the M_2 period.[32] He estimated the natural period of the Bay at 9 hours and 3 minutes, supporting earlier results of Redfield[33] and Ippen and Harleman.[34] Yuen[35] investigated the effects imposed by barrier construction in the Fundy system and found that, in general, the tidal amplitude would decrease approximately 20 percent and the time of high water would advance. His two-dimensional model encompassed the entire Bay, out to the vicinity of Passamaquoddy Bay. Other models of the effects of barriers on the Bay were expanded to include part of the Gulf of Maine. Duff,[36] calculating the reduction in the M_2 tide for the case of a hydro-electric generating barrier at Cape Enrage, for the first time included the Gulf of Maine as a critical consideration in Bay of Fundy resonance studies.

Present understanding of the resonant tides in the Bay-Gulf system stems from Garrett's research[37] examining the possible resonance of the Gulf and Bay with the M_2 tide. He compared the ratios of the major semi-diurnal tide characteristics within the Bay-Gulf system to those outside of the region along the East coast of the North American continent to Bermuda. The responses of the system were analyzed against the forcing characteristics of the ocean tide. The consistency of the tidal ratios for various tidal stations suggested that the Bay's response was integral with that of the Gulf. Garrett's estimate[38] of the resonant period for the system was 13 hours and 18 minutes, close enough to the forcing period of the M_2 tide to explain the resonant response. Garrett[39] further investigated a normal mode model of the Bay-Gulf system.

Table 2. **Comparison of Gulf of Maine-Bay of Fundy Studies**[30]

	Area of Model	Model Type	Grid Size	M_2 Period	Barrier location and type	Changes in Regime
Proudman	BOF	1-dimension	NA	11.06	NA	NA
Rao	BOF	1-dimension	NA	9.047	NA	NA
Yuen	BOF	2-dimension	8 Km	NA	Cape Blomidon impermeable	15% amplitude decrease & 20' phase decrease
Greenberg	BOF	1-dimension	17 Km	9.05	Cape Blomidon impermeable	26% amplitude decrease
Duff	BOF/GOM	2-dimension	17 Km	theorized near resonance condition	Cape Enrage impermeable	34% reduction
Parkinson	BOF	2-dimension	5 to 16 Km	NA	Economy Point permeable	9% amplitude decrease
Garrett	BOF/GOM	2-dimension	NA	13.3	NA	NA
Garrett	BOF/GOM	2-dimension	25 Km	12.981	Cape Blomidon permeable	7% decrease at barrier 16% increase at Yarmouth 25% increase at Boston
Greenberg	BOF/GOM	2-dimension	varied from 21.14 Km to . 1.61 Km	12.5	Economy Point permeable	6% decrease at barrier 5% increase at Yarmouth 9% increase at Boston

In this case, he estimated that the normal mode would be between 12 hours and 23 minutes and 12 hours and 59 minutes. Greenberg's refinements[40] place it at 12.5 hours (Figure 6).

Greenberg's Model

Greenberg's numerical model explains the effects of tidal barriers on the Bay of Fundy - Gulf of Maine.[41] Building on past modeling mentioned above, it expands the system boundary to the edge of the continental shelf. The model also provides estimates of the resonant period of the Bay-Gulf system and simulates the effects of changing barrier locations on tidal amplitudes and phases.

The model considers the region as separated into a grid of boxes, each acting as a small parcel of fluid (Figure 7). The forces on each parcel are resolved for each time step, and the future state of the element is predicted. Since each element affects neighboring elements, and in turn is affected by them, the routine begins at the boundaries and proceeds toward interior regions. The results of the model depend on the grid size employed, the time intervals used, the approximations applied in calculating the forces on each box, the acceleration of the fluid, and the nature of the boundary conditions.

The 1979 model incorporated four grid sizes: 21.14 km grids for the Gulf of Maine from the continental shelf edge to the upper Bay of Fundy; 7.047 km grids covering the central part of the Bay; a fine grid of 2.349 km covering Chignecto Bay and the Bay of Fundy; and an even smaller grid size of 1.61 km for Minas Channel and Minas Basin. Grid sizes are selected to provide the resolution needed in the regions of interest where the fluid properties are expected to change rapidly, i.e., finer grids for better resolution at barrier location. Tidal actions within each grid can then be represented mathematically, and their effect on each subsequent grid calculated. An integration of the tidal response through each cube results in a model of the entire system.

The fluid in a particular grid box has a number of forces exerted upon it: the difference in elevation of the fluid in the neighboring boxes; the friction acting primarily at the surface and bottom of the sediment; the force of gravity; and the Coriolis force, resulting from the rotation of the earth. Boxes bordering the system exhibit the effects of frictional forces on their sides as well. From these forces and a previously known velocity as initial conditions, one can calculate the acceleration of the fluid within each box. The resulting velocity field is then used to calculate the extent to which a particular box is filling up or emptying and the change in elevation of the fluid in the box. These results are then employed to calculate a new elevation, and the process is repeated for each fluid parcel and for each new time step.

In the above calculation, the hydrodynamic equations of motion and continuity relate the basic principles involved for shallow water waves (a progressive gravity water wave whose depth is much less than the wave length). The underlying principle

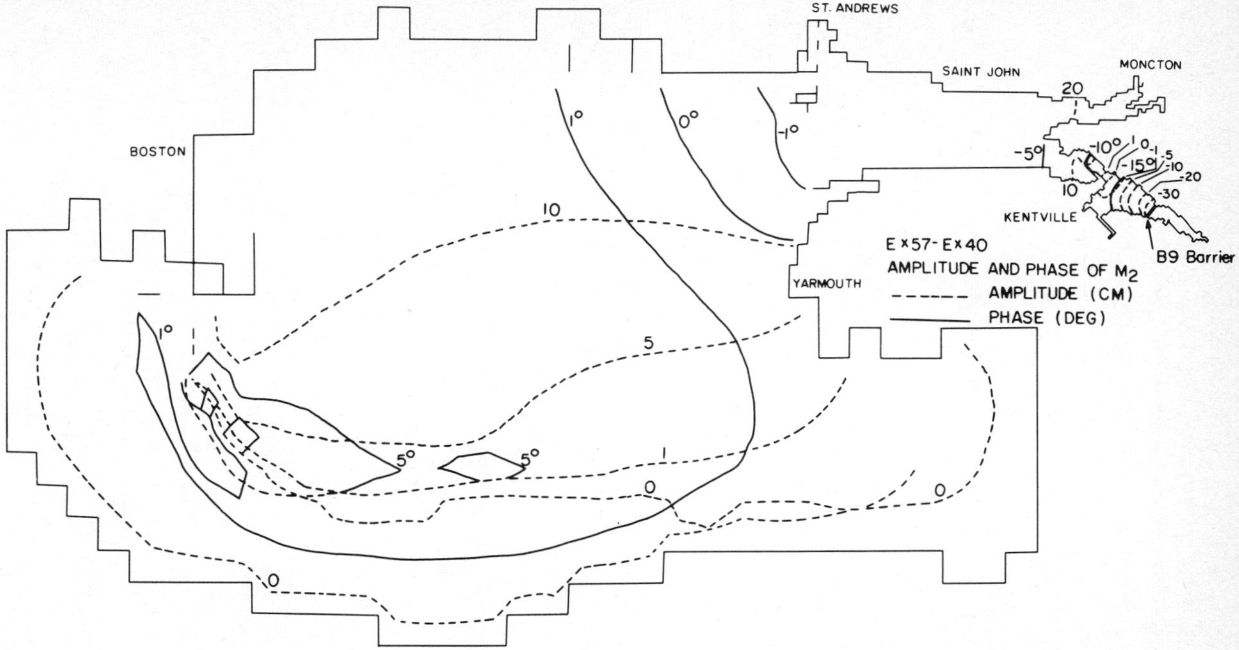

Figure 6. Change in Tidal Regime Due to Barrier at B9 Site.

Used with permission of Dr. David Greenberg, Bedford Institute of Oceanography, Dartmouth, Nova Scotia.

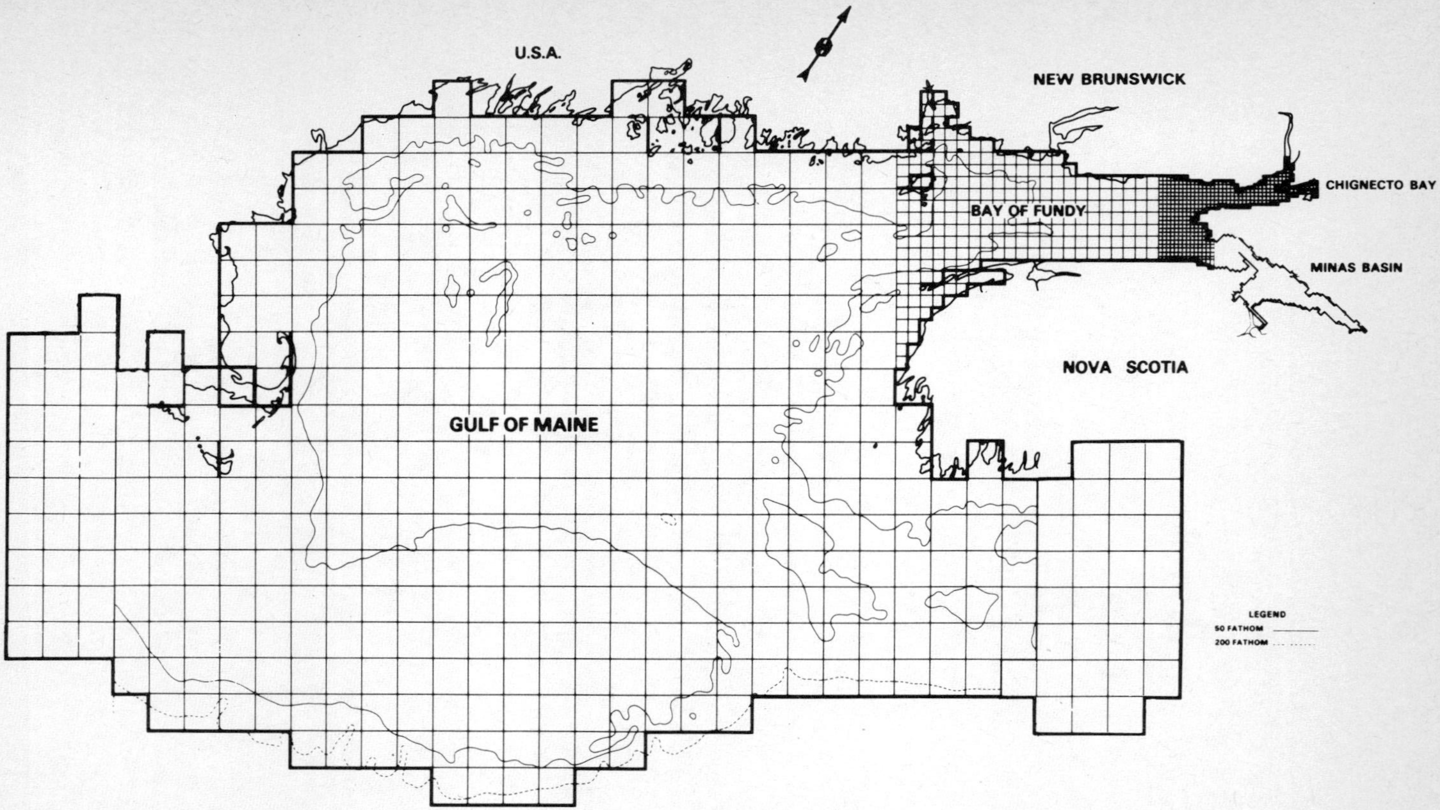

Figure 7. Grid System Employed in the Greenberg Model.

Used with permission of Dr. David Greenberg, Bedford Institute of Oceanography, Dartmouth, Nova Scotia.

for the equation of motion is given by: mass x particle acceleration = friction + gravity (a constant) + pressure gradiate + Coriolis force. This equation is applied to a continuous volume of fluid. Consequently, the equations must account for the fact that different parts of the fluid are moving in different directions at different times and places, and that no two fluid parcels can occupy the same part of a space at the same time. These conditions are satisfied through the continuity equation.

Friction is a necessary component in the calculation of acceleration. In hydrodynamic systems like the Bay and Gulf, the magnitude of the frictional force is essentially determined by turbulence (the motion of a fluid in which local velocities and pressures fluctuate in a random manner). No exact methods exist for calculating the effects of turbulence. Semi-empirical methods, however, approximate these although they become less and less reliable as the situation becomes more complex.

Frictional dissipation opposes the direction of currents and is proportional to the force of the current. Frictional forces are also exerted at the water's surface by the wind because the water is slightly viscous, and viscous forces oppose the flow of the water. With the wind blowing, the frictional stresses applied at the surface will be transmitted downward. If the wind stops, water movement will begin to slow down, and eventually will stop as viscosity acts to transfer the kinetic energy to heat energy. The dynamic effect of friction then is to redirect partially the currents on shoals to deeper water where dissipation decreases.

The primary effect of gravity in shallow water tidal motion is in the determination of the pressure gradient force. The pressure gradient force accelerates a particle towards the region of lowest pressure. Acceleration is proportional to the gradient, and the density of the water is assumed constant.

The Coriolis force accounts for the fact that a water molecule accelerated by the forces of friction, gravity and pressure is also accelerated by the earth's rotation with respect to an inertial reference frame. The acceleration is proportional to the particle velocity relative to the earth. In the northern hemisphere, it is directed toward the right of the direction of fluid motion. The magnitude of the force is given by: $2 \Omega \sin \Phi v = fv$, with $f = 2 \Omega \sin \Phi$, Ω the angular rotation rate of the earth, Φ the latitude, and v the magnitude of the velocity. Since this effect increases with latitude, Greenberg's model assigns different values to f for each grid size.

Besides accounting for the physical forces involved, it is also important to include the effects of the fluid boundary and the coastline. It is necessary, in other words, to ensure that the liquid does not flow through the walls of its container. If fluid enters a grid through an opening, then this flow is treated as a source; if fluid leaves, the flow out is treated as a sink. Sink effects, requiring incorporation into the model scheme, include mud flats and sand bars, which become very shallow or even dry out when the water level becomes low. In addition, the coast is often indented by coves or has back eddies behind rocks to be considered.

Greenberg's model combines explicit and implicit computations. Equations to be solved are the vertically integrated equations of momentum and mass conservation. These shallow water wave equations are written in finite difference form, a calculus procedure whereby the difference between the values of a function at two discrete points is used to approximate the derivative of the function. The resulting system of algebraic equations, as noted, is solved on a step-by-step basis for each grid box and then by numerically summing them to obtain a model of the tidal movements. These numerical equations, however, hide a great many things. Instead of having continuous variations in time and space, the variations come in steps. By making the steps small enough, one can overcome the difficulties arising from this, but making the steps very small greatly increases the expense of the calculations. This model compromises the two extremes by using a successively finer grid as the Bay of Fundy shallows and using the large grid for the Gulf of Maine.

Procedurally, the model was developed for the Bay-Gulf region subject to the boundary conditions of zero mass flow across the coastline and a prescribed tidal elevation at the open boundary. The open boundary is taken to run along the edge of the continental shelf where the model adjoins the open ocean (See Figure 3). Prescribed inputs at the open boundary are derived from direct observations, i.e., tidal station records. Greenberg used observed data to adjust the model until it reproduced the tides observed at the coast. Several tidal models of this sort have been constructed with considerable success.[42] A major factor in this success is the use of empirical data on tidal heights and to a lesser extent on tidal currents. The use of such data obviously improves the accuracy of the semi-empirical parameters. Once the model met the observed situation, it was rerun, with the tidal barriers inserted at prescribed locations.

The numerical procedures also included simulations of the drying and flooding of shallow areas. Advective (mixing) terms were determined by observing areas which had characteristics of large currents that rapidly changed their direction in confined areas. These parameters were included in the Minas Basin and Minas Channel equations.

Initially, the model was run with the M_2 tidal response specified at the grid points along the edge of the open boundary. The M_2 wave was then correlated with the records of the M_2 tide from offshore tidal gauges. The results for the M_2 response were examined by analysis of one tidal cycle. Adjustments included varying the friction parameters in various areas as well as the open boundary elevations to give the best agreement with shore-based tidal stations (Table 3). Along with these changes, the calibration required an increase in the depth of the Gulf by 3.6 meters. This increase was justified because the depths of the water in the larger grids of the Gulf were chosen by averaging individual depth soundings recorded on hydrographic charts (usually less than tens of values for each grid square). These charts primarily show minimum depths for shipping, thus many deeper spots between the lows are ignored. The target accuracy for the model of within 15 cm in amplitude and 5 degrees in phase with the shore-based data was satisfied.

Table 3. Comparison of M_2 Amplitudes and Phases Obtained from Tide Gauge Observations and from Greenberg's Numerical Model.[43] For each port, the record length of the observations, on the basis of which M_2 was determined is given.

Name	Amplitude (cm) Obs	Amplitude (cm) Model	Phase (deg) Obs	Phase (deg) Model	Record Length
Cape Cod Lighthouse	116	108	356	356	29 days
Cape Cod Canal (east)	134	138	355	355	58 days
Boston	134	133	354	352	5 years
Portsmouth	121	129	354	349	1 year
Portland	133	134	347	345	5 years
Rockland	146	147	342	340	1 year
Bar Harbor	155	160	339	335	1 year
Eastport	261	267	346	338	3 years
Back Bay	254	260	340	337	29 days
St. John	309	301	342	340	537 days
Herring Cove	406	425	348	348	26 days
Cape Enrage	436	438	350	348	2 x 29 days
Grindstone Island	472	468	351	351	2 x 29 days
Hopewell Cape	496	476	351	353	2 x 29 days
Belliveau Village	555	492	358	357	15 days
Joggins Wharf	488	476	351	352	29 days
Cape Capstan	413	422	347	340	29 days
West Advocate	409	419	342	347	29 days
Cap D'Or	434	432	346	348	197 days
Port Greville	461	464	356	353	29 days
Dilligent River	488	481	357	356	84 days
Cape Sharp	480	499	5	359	29 days
Parrsborough	505	519	5	5	29 days
Five Islands	542	542	9	14	64 days
Burntcoat Head	564	553	14	25	29 days
Hantsport	567	554	14	15	29 days
Cape Blomidon	503	532	4	6	44 days
Scots Bay	467	468	349	350	29 days
Margaretsville	386	388	337	339	62 days
Parkers Cove	332	330	332	333	50 days
Deep Cove	309	299	330	330	56 days
Centreville	265	269	334	329	113 days
West Narrows	223	218	322	321	81 days
Westport	214	212	322	322	57 days
Mateghan	203	203	316	317	73 days
Port Maitland	185	185	310	313	29 days
Yarmouth	168	160	309	307	1 year
Pinkney Point	139	152	304	304	116 days
Abbott Harbour	133	130	276	286	47 days
Seal Island	120	124	296	288	44 days
Clarks Harbour	109	112	272	265	82 days
Lockeport	70	70	243	243	29 days

Additional support for the model was provided by the correlation of the kinetic energy and potential energy in the system and also the correlation of work (the transfer of energy to a body by the application of a force over a distance) done by the input tide and the dissipation due to friction. The rate of energy change contained in an oscillating mass of water in a basin on a rotating earth is written out using the equations of hydrodynamics. The concept of power transfer, thus introduced, enables one to follow the transit of power from one part of the basin to another. It is then possible to establish a balance of power present in the basin.

Work done at any one instant in time is positive if the current is flowing into the system and the elevation is positive or if the current is flowing out of the system with elevation negative. The rate of work done by the pressure forces over the surface bounding a given volume of fluid equals the total rate of change of the kinetic and potential energy inside the fluid plus the work done by the tidal and frictional forces.

Some energy enters the Bay-Gulf system in the form of a tide wave. According to the law of conservation of energy, the amount of energy pushed into a system (i.e., by a tidally generated wave) must equal the amount of energy leaving (i.e., through dissipation) or vice versa. The values derived, computed by Greenberg, show the tidal regime produced by the model to be in balance. In the model, mean total dissipation, 5.50 x 10-10W over one tidal cycle, closely balanced the mean rate of work done across the open boundary, which was found to be 5.41 x 10-10W. Thus, the energy balance coupled with the close correlation of the actual tidal observations in the system to those of the modelled tide provides sufficient support to conclude that the model as constructed by Greenberg is reasonably accurate.

The Tidal Regime Produced by the Model

Potential energy is highest in the upper reaches of the Bay where the tidal range is greatest. In the Gulf of Maine (exclusive of the Bay of Fundy), the kinetic energy is greater than the potential energy, because the energy is integrated over a large area with minimal storage of energy. Computational values of frictional dissipation show it to be concentrated in a wide band around the Nantucket Shoals, through Georges Bank, across the Fundian Channel and into the Bay of Fundy, with maximum values over the shallow areas. Very little dissipation occurs in the vicinity of New Hampshire where mean kinetic energy is lowest. Frictional dissipation gradually increases as the depths in the Bay decrease, particularly into Chignecto Bay.

The principal flow of net power is into the Minas Basin. The net power transferred across the open boundary of the model is largest at the entrance of the Fundian (Northeast) Channel. Within the Gulf of Maine, the net power transfer divides into the counterclockwise gyre, following the U.S. and Canadian coastlines and the near-shore side of Georges Bank, and into a branch directed into the Bay of Fundy.

The model also indicates the resonance period. The amplitudes and phases as specified for M_2 and the open boundary were used, and the period of the input tide varied. The resonance period was found to be about 12.5 hours. Insertion of barriers in the Bay of Fundy would shorten the system, bringing the natural period of the system closer to the M_2 period of 12.42 hours.

Effects of Barriers

The Greenberg model derived tidal amplitudes and phases for barriers at the A6 (Shepody Bay) and B9 (Cobequid Bay) sites (See Figure 2 in chapter 1). Permeable barriers simulating ebb-flow power generation were included.[44] Simulation accounted for reservoir filling and emptying behind the barrier, discharge of water through turbines, and periods of no flow through the turbines, relating to the design features of the proposed systems. The operation of the scheme involved periods of no flow through the barrage as well as others of seaward discharge.

Barrier lines chosen coincided with the boundaries of the grid squares. The sites examined for barrier location enclosed Shepody Bay (A6) and the upper end of Minas Basin (B9). Specific flow parameters were used for each barrier site. The Minas Basin barrier (B9) produced a general increase in tidal amplitude throughout the Gulf of Maine with a drop in amplitude at the barrier site and nearby, but an increase in the rest of the Bay of Fundy (Figure 8). Effects on the tidal regime would be observed as an earlier arrival of the tide at the barrier site and later arrival in the Gulf. Changes produced by a barrier at the Shepody Bay location (A6) were lower in magnitude, but generally showed the same effects in the Bay and Gulf as did the Minas Basin enclosure. Barrier placement reduced the period of the Bay of Fundy alone, decreasing the tidal amplification from the mouth to the head of the Bay. Decreased flows through the Minas Channel, and into Minas Basin, however, resulted in correspondingly decreasing frictional dissipation, allowing an increase in tidal amplitudes.

Greenberg's model predicts tidal regime changes at various ports in the Bay and Gulf (Table 3). The Shepody Bay barrier produces the greatest change directly in front of the barrier where a decrease of 24 cm in range is observed. The resulting tide would be 24 cm lower at high tide and 24 cm higher at low tide. Other changes observed due to the barrier occur in the lowest reaches of the Gulf of Maine. The tide is 4 cm higher at high tide and 4 cm lower at low tide from Portland to Boston. Phase changes are on the order of 2 to 4 minutes earlier. A decrease in phase lag is equivalent to an advance in the time of high water.

The changes due to the Minas Basin enclosure (B9) are approximately 3 times that of the Shepody Bay barrier (Table 4). Again, the greatest change is found directly in front of the barrier. A 34 cm decrease in the high tide and a 34 cm increase in the low tide are predicted. In the lower reaches of the Gulf, the range of the tides is increased. The high tide at Boston is predicted to increase by 15 cm and that of the low tide to decrease by 15 cm. At Portland Harbor, high and low tide levels change by 14 cm as they would at Digby, Nova Scotia. A change of only 9 cm at ebb and flood times is predicted to occur at Yarmouth, Nova Scotia. Greatest increases occur in St. John, New Brunswick, where high tide would increase by 18 cm and low tide by the same amount. The phase changes are roughly 2 to 4 minutes later in the Gulf and 4 to 6 minutes earlier in the Bay of Fundy.

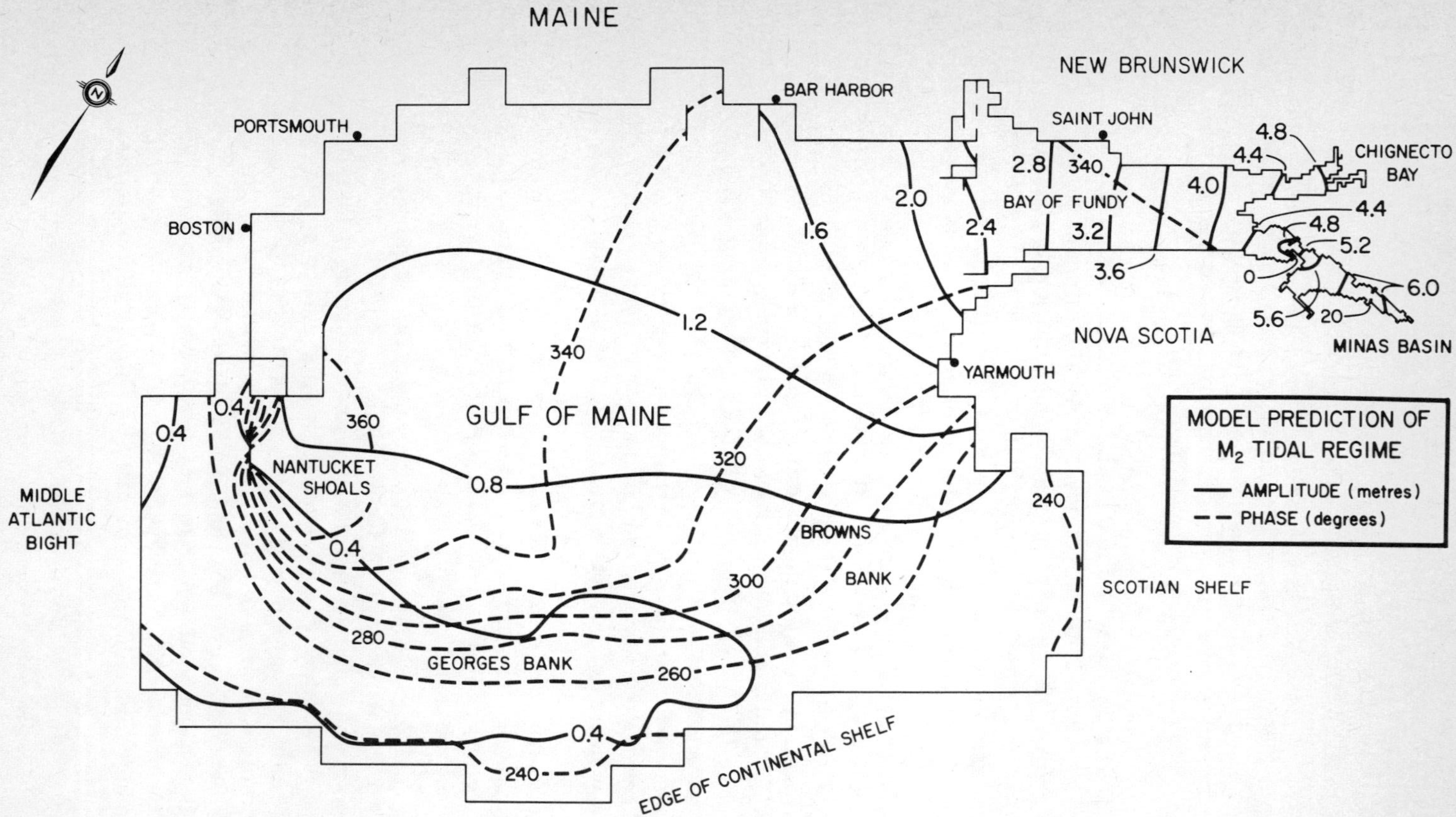

Figure 8. M_2 Tidal Regime Predicted by Greenberg Model.

Used with permission of Dr. David Greenberg, Bedford Institute of Oceanography, Dartmouth, Nova Scotia.

Table 4. Amplitudes (in centimeters) and Phases of M_2 Locations Within the Model, Changes from the Natural Regime as Determined in the Model, and Difference between Mean Reservoir and Mean Sea Level for Barrier Cases A6 and B9.

	A6		B9	
St. John	304	339°	321	338°
Diff.	+1	-2°	+18	-3°
Digby	306	330°	319	330°
	+1	-2°	+14	-2°
Yarmouth	163	309°	170	311°
	+2	-1°	+9	+1°
Portland	138	346°	148	349°
	+4	-1°	+14	+2°
Boston	139	353°	150	356°
	+4	-1°	+15	+2°
In front of Barrier	454	349°	558	358°
	-24	-1°	-34	-17°
Behind Barrier	162	57°	199	67°
Mean Level Diff. Across Barrier	287		337	

Tides show considerable natural variation because of the number of forces acting on them such as meteorological conditions and coastal configuration. For example, the average daily tides at Boston Harbor, vary between 9 to 10 feet in range.[43] Over a year, that range may extend from 7 to 13 feet. Further, an 8 percent variation in tides arises because of an 18.6 year tidal cycle. The changes predicted by the Greenberg model then would increase the yearly ranges to 7-1/2 feet and 13-1/2 feet. Storms could increase these values as well as reduce their effects. The model also predicts currents resulting from the tidal modulations. Up to a 5 percent shift may occur, and this shift could effect mixing throughout the Bay and the Gulf. Care, however, must be taken in the interpretation of residual (mean) circulation patterns produced by the model. No tests have been conducted to test its reproduction of such second order effects. Advection terms were only included for the Minas area and here residual currents compare well with observations.

Nature of Effects

Barrier emplacement will result in energy that is normally dissipated upstream of the barrier site to be redistributed downstream of that site. Without the barrier, the tidal wave propagates to the upper reaches of the estuary where the wave is normally partially reflected as well as deflected by bottom stresses. With the barrier in place, however, the tidal wave would be reflected at the barrier, changing the tidal regime locally and in more distant areas. Potential effects on transboundary resources will be either direct, for example turbine-induced fish mortality, or indirect, resulting in secondary effects that may accelerate coastal erosion, change patterns of water mass mixing, and alter nutrient fluxes within and between the Bay and the Gulf.

It may be possible to reduce uncertainty about direct effects by examining the effects of analogous projects and from experimentation. Causality is relatively easy to establish. Indirect effects, however, represent a more serious problem because of the role that interactive factors, blocking or magnifying events in ways usually difficult to predict, play in structuring a complex ecosystem. The indirect environmental effects speculated to occur from the tidal power project represent an intensification of existing natural forces, that is, heightened tides and the consequent increased mixing of water masses. Absent an understanding of the specific behavior of the interactive factors, predicting their direction of change and magnitude after dam emplacement is extremely difficult.

Among the factors that may block or amplify Fundy tidal power's effects are those which affect tidal energy dissipation, including bottom topography and coastal configuration, astronomical cycles, and meteorology. Most of the changes suggested, however, will probably only have a short-term effect because the Bay of Fundy-Gulf of Maine ecosystem has throughout geologic time experienced sea level fluctuations with the range of magnitude suggested for the tidal power facility. Some temporary economic dislocation may result as coastal and open ocean systems adjust to the changed distribution of energy.

It is important to remember that the predicted change is in tidal amplitude - higher high and lower low tides - not in mean sea level. Mean sea level is rising along the northern New England coast at a rate of approximately 0.3 cm per year. Thus, any changes in tidal amplitude from Fundy tidal power would be imposed on a rising mean sea level, although the latter would not necessarily magnify the effects of the former.

Direct Effects

Fish Mortality. Fish swimming through the tidal power facility's turbine may be subject to mutilation from turbine blades. Among the few commercially exploitable species found in the upper Fundy, the Atlantic shad (*Alosa sapidissima*) would be among the most susceptible. Populations of this species annually migrate into the Bay of Fundy from as far south as the St. John's River in Florida,[45] and they are fished along the entire eastern seaboard.

Shad enter the Bay late in May, traveling in the counterclockwise residual current. Highest densities occur from late June until late July in the Minas Basin and slightly later in the Cumberland Basin. The fish may remain up to twenty days in the Minas Basin and six in the Cumberland. Only limited tag return studies have been conducted on the Fundy, but they do suggest that at a minimum 30 percent of the Minas-Cumberland shad population, calculated to be around 6 million adults, come from U.S. rivers.[46]

While in the upper Fundy, shad travel passively with the tides near the surface as they feed on the abundant mysid shrimp. It is estimated that 1×10^5 shad move with each tidal cycle.[47] Over their twenty day residence in the Minas region, the tides will carry them past a given point multiple times. Thus, a large portion of the population would be expected to swim through the turbines.

Dadswell et al. have calculated estimated potential fish mutilation rates (Table 5) for some upper Fundy fish, including shad, using Van Raben relationships.[48] Population densities were based on those expected at the Minas site. These relationships allow determination of impact velocities (fish against the blade), the distance between each pass of the blade, and resulting fish mortalities. Turbines like the Straflo (the type planned for the Minas site and currently installed at Annapolis) present little hazard of mutilation to small fish or medium-sized fish passing through once in a downstream migration. The large blades are slower moving than those of other commercial turbines - Kaplan or Frances. Impact velocities are greater, however, because the outer end of the large diameter blade is moving more rapidly than are those of comparable turbines. Thus, larger resident fish - striped bass and Atlantic sturgeon - as well as fish making multiple passes - shad and possibly salmon - could experience a greater potential for mutilation.

Table 5. Potential Mutilation Rates* for Upper Bay of Fundy Fish Species[49]

Species	Length	
Alewive	30 cm	5.7%
Shad	50 cm	9.5%
Salmon	80 cm	15.2%
Bass, Dogfish	100 cm	19.0%
Sturgeon	100 cm	38.0%

*mutilation rate = (fish length x mutilation ratio (theoretical/observed))/(water length).

Survival may be calculated using an exponential relationship where the original population, N, times 1 minus the mutilation rate equals survivorship. For a population starting at 100,000 after 5 passes (100,000 (1-0.095(5)), 60,700 will survive. After 10 passes, 36,800 will not be hit, and after 20, 13,500 will be left. Experimental tests of Van Raben relationships suggest that the relationships overestimate mutilation rates for 5, 10, and 20 passes so that the above numbers should be doubled. Mutilation rates then then become 20%, 35%, and 45%.[50]

Fishways, like the one installed at Annapolis, represent one way to circumvent turbine blade mortality. However, at Annapolis fish are on a spawning run and the dam has reopened an area previously closed by a causeway.[51] Thus, the Annapolis structure will allow evaluation of the response of shad to the fishway rather than indicating rates of mutilation for multiple passes.

Transboundary fishery resources may also be susceptible to habitat changes occurring as the direct result of dam construction. Some migratory fish such as the shad are attracted to the upper Fundy because of the high concentration of food in its turbid waters. The barrier, by deflecting tidal energy, will reduce this turbidity,[52] and possibly reduce the invertebrates upon which the migratory fish feed. This habitat change may result in higher concentrations of these fish in adjacent areas, for example, the Cumberland Basin. Further, these species may find the changed environment less hospitable and so counteract the effects of turbine mortality. The time necessary for these species to adapt to the changed conditions, if indeed they do, is unknown.

Migratory Waterfowl. The tidal flats of the upper Fundy represent an important feeding ground on the Atlantic flyway from the Arctic to South America for many migratory shorebirds. Formation of a headpond behind the tidal barrier will destroy almost half of the intertidal zone, an important foraging area for millions of these birds.[53] Not only would fewer food items be available but fowl would also devote more time competing for the scarce resources. Recent studies suggest though that some species thought to feed only on a limited suite of intertidal invertebrates also exploit species associated with wrack along rocky coastlines.[54]

Indirect Effects

Coastal Erosion. Higher high tides and lower low tides will shift patterns of sediment movement along the Bay-Gulf coastline. They will potentially affect beach erosion, inlet stability, shallow water nutrient fluxes and extent, and faunal diversity of salt marshes. Storm action, particularly from storm surges, will contribute to these alterations, although the increment attributable to Fundy tidal power will be difficult to document.

The most damaging coastal effects of storms result from storm surges. These result when strong winds blowing toward the land create frictional stress. The water "sets up", causing an increase in water level as much as several meters.[55] When associated with cyclonic storm conditions, storm surges may be intensified by decreases in atmospheric pressure. For example, a 3 kPa (= 30 millibars) drop may cause a 0.3 meter rise in sea level.[56] Three typical storms tracked in New England caused a 6 kPa (= 60 millibars) enhancement of nontidal sea level which translated into increases of three to four feet.[57] Storm surges attacking the New England coast during the most recent 100 year storm, the blizzard of 1978, exceeded one meter.[58]

Coastal effects of storm surges may appear as much as five or six hours after peak winds, but the variance may be one to twenty hours.[59] Topography and speed of storm movement amplify or diminish these effects. For example, the islands and promontories around Portland, Maine protect it from most storm effects except for those from the southeast.[60] Timing of high tides and storm surges is also an important factor. The most dramatic effects of a storm on an amplified tidal regime would be expected during a high tide. U.S. Army Corps of Engineers data for New England storms indicate that storm surges have rarely correlated with high tides.[61]

A 30 cm change in tidal range could result in storms currently ranked as 100 year storms (waves 3.10 meters above sea level) being classified as the more frequent fifty year storms (waves 2.97 meters above sea level).[62] In turn, fifty year storms would be designated as thirty year events. Direction of storm movement and local topographic conditions will influence the intensity of these storms. Barrier construction should not affect either the absolute number or severity of storms impacting the coast.

Wind direction, duration of the storm, and coincidence of the storm with any given tidal height (high, low, or slack) are crucial but largely unpredictable factors in any determination of storm impact. The roughly 10 percent increase in tidal amplitude predicted by the Greenberg model when matched with these storm-related factors suggests that coastal storm damage could also increase by approximately 10 percent. The mitigating forces mentioned above would, however, probably work to minimize this figure.

Accelerated erosion of sand beaches over the long term is a matter of particular concern. Sand beaches in Maine are characteristically low energy environments with little or no alongshore transport and susceptible to wave attack primarily from a southeasterly direction.[63] Discrete, high energy events such as storms alter the normal wave refraction pattern and result in the transport of sand within the boundaries of the system but in an offshore direction. As the event passes and these areas return to equilibrium, the net accretionary tendencies of these sand areas gradually restore their original profile by returning to shore the sand previously eroded. Consequently, these intermittent, high energy events change only temporarily the beach's profile and the net onshore transport of sand from within the essentially closed and sediment-limited system. Rarely is there any permanent net loss of sand.

Since the predicted changes in sea level from barrier construction concern only tidal amplitude changes and not mean sea level shifts, these amplitude changes would probably have little effect on the sandy foreshore areas of Maine's beaches. Erosion of backshore areas, where the dominant transport agent is wind, would not be greatly altered either by tidal amplitude changes.

In contrast, stabilized beaches altered by human intervention create another problem. These beaches are inherently unstable, and any change in energy regime such as a storm would force a redistribution of the eroding sediment by increasing alongshore transport in the direction of lower energy environments. In these locations, for example around Wells, Maine, and Plum Island in Newburyport, Massachusetts, any storm event may significantly affect the shoreline regardless of barrier emplacement.

An alternate view exists on this issue. Some argue that a 10 percent increase in amplitude can be translated into a 10 percent increase in net beach erosion.[64] Following this reasoning, an estimate of a 20 meter retreat was calculated for Old Orchard Beach in southern Maine. This must, however, be tempered by an important consideration. The derivation applied was developed for the case of mean sea level change, not tidal amplitude shift - the case if barriers are emplaced. The authors argue that the tidal range increase will heighten the elevation of wave energy input to Maine's sand beaches just as mean sea level increases do. Sea level changes affect beach profiles, particularly offshore, differently than do tidal amplitude changes. Thus, the 20 meter estimate should be viewed with considerable caution.

Biota. *(a.) Intertidal Fauna:* Translating a tidal increase into predictions of changes in intertidal fauna is difficult because intertidal areas already experience daily, seasonal, and long term variations in sea level and tidal range. Intertidal communities have also exhibited considerable resilience to changes in physical parameters. Those sometimes exposed to saline conditions would experience those conditions for longer periods. Perhaps the only organisms likely to experience any real impact are those currently not exposed to saline conditions.

(b.) Salt Marshes: Tidal salt marshes may also be vulnerable to changes in tidal amplitude. Marshes along the northern New England coast are shallow features, extending to a depth of only about five meters.[65] Little knowledge, however, exists on their growth and development. Increased inundation would doubtless shift faunal distributions, perhaps even resulting in a greater abundance of the more productive low marsh cord grass, *Spartina alterniflora*. Debate exists whether the region's marshes are accreting or eroding.[66] Thus, it is difficult to predict what effect increased flooding would have on marsh elevations. Expanded radiocarbon dating of the underlying peat would give some indication of their direction of change. Without this understanding, it is difficult to hypothesize the effects of tidal power.

Tidal Mixing. Increasing tidal ranges will also enhance mixing in coastal waters, particularly at the mouth of the Bay of Fundy around Grand Manan Island, along the coast of southwest Nova Scotia, and along the northern edge of Georges Bank. Each of these areas exhibits high fisheries productivity as a result of nutrient rich waters being aerated by the turbulent mixing in shallow waters. The degree to which heightened mixing will contribute to increased productivity is unclear. While many species appear to be closely tied to narrow environmental ranges, the extent of those ranges is unknown.

Herring, for example, utilize the complex hydrodynamics at the mouth of the Bay of Fundy for the aggregation and dispersal associated with their breeding, spawning, and feeding.[67] Principal herring spawning areas - southwest Nova Scotia, Grand Manan, Georges Bank, and Nantucket Shoals - lie generally within vertically well-mixed areas. These are bounded by zones where the water column changes rapidly from well-mixed to stratified.[68] At Grand Manan and Georges Bank, residual circulation takes the form of tidally-induced gyres. Associated with these gyres are areas that are more completely vertically well-mixed.[69] These areas are defined and delimited by well-mixed/stratified transition zones that provide a physical, stable retention mechanism.[70]

Fundy herring populations spawn in the fall, and larvae overwinter in the lower Bay. Despite their relative lack of motility, evidence suggests that the larval herring use selective vertical migrations in these complex areas to be "retained" in suitable nursery areas. Tidal and residual circulation patterns in the lower Bay and the mixed/stratified areas they produce seem to provide a stable geographic mechanism that is essential to the life history of several important species.

Changes in the factors that determine whether the water column is mixed or stratified could alter this physical retention mechanism and affect the distribution and abundance of both an important marine food source - icthyoplankton like the euphausiid crustaceans - and commercially valuable transboundary fish species like herring.

Garrett, Keeley, and Greenberg have tested a criterion for tentatively distinguishing between well-mixed and stratified waters.[71] These authors applied a mathematical relationship developed by Simpson and Hunter[72] to tidal regime changes predicted by Greenberg's 1978 tidal model. The results suggest that a barrier emplacement would cause areas near Great South Channel and Georges Bank to change from stratified to well-mixed.[73] The predicted changes could affect local productivity either by increasing nutrient supply to surface layers, or by integrating the surface phytoplankton production into the unstratified water column, thereby decreasing phytoplankton production and depressing local production overall.

Scallops also exhibit behavior closely attuned to mixing and circulation. Sea scallops (*Placopecten magellanicus*) occur in beds 30 to 40 fathoms deep on the gravel and boulder bottom along the eastern edge of the Bay of Fundy, near Digby, Nova Scotia. Observations over a fifty-year period suggest that water circulation and water temperature are related to year class strength.[74] Strongest year classes appear to correlate with spawning temperatures between 10 degrees C and 11.5 degrees C.[75] No scallops in these beds appear to spawn at water temperatures below 9.5 degrees C.[76]

Water circulation and temperature control larval settlement. Scallop larvae, like plankton, travel in the counterclockwise circulation of the Bay for four to six weeks prior to settlement.[77] A counterclockwise gyre variably forms at the mouth of the Bay and may return larvae to the beds where they were spawned. Higher water temperature stimulates more rapid larval development and therefore earlier settlement than does colder water.[78]

Tidal mixing, insulation, and water exchanges between the Bay and the Gulf control water temperature in this region. Increased mixing and reduced exchanges will elevate water temperature. When the gyre closes the Bay's circulation, larvae are deflected back to the Digby beds, and water temperature rises. Larvae, therefore, stand a better chance of settlement in the Digby region. Colder temperatures pospone settlement and increase the possibility of larval predation. More open circulation also results in larval transport outside the Fundy region. Scattered beds consisting of one year class probably result from larval dispersal during colder years.[79]

Caddy identifies nine year cycles associated with the abundance of Digby scallops and suggests that these correlate with the lunar apsides (8 year cycle) and the nodal cycle (18.6 years).[80] These cycles enhance tidal mixing, potentially warming surrounding waters. Discussion of the transport of larval scallops, however, is largely speculative because larvae have yet to be found in plankton tows.

Dickie's and Caddy's hypotheses suggest that temperature increases of only a few degrees enhance larval survival.[81] A shift in tidal amplitude would probably raise water temperatures in the Digby area to the benefit of the local scallop population. Altered tidal mixing might also affect the strength of the Grand Manan gyre, but the direction of this change and its influence on larval scallops are unclear. Insufficient information exists on scallop populations along Maine's coast to speculate about the effects on them of a shift in tidal mixing.[82]

Intensified mixing in some regions of the Gulf of Maine has been suggested to stimulate the production of the red tide organism, *Gonyaulax excavata*, which causes paralytic shellfish poisoning.[83] Extensive concentrations of *Gonyaulax's* resting stage coincide with the area of the Grand Manan gyre at the mouth of the Bay of Fundy.[84] Seasonal changes in the gyre's intensity may explain the periods when *Gonyaulax* are broadcast to the south and west along the New Brunswick and Maine coasts. While changes in mixing as a result of Fundy tidal power may affect the intensity of the gyre, it is unclear to what extent and what the implications would be for the red tide organisms, particularly in light of the scarcity of information on their behavior.

Sea Level Changes. Dicussions about the effects of Fundy tidal power sometimes confuse changes in mean sea level with the changes in tidal range which may result from dam emplacement. Sea level in northern New England and maritime Canada has been rising for the past 8,500 years. This increase results from the complex interaction of crustal responses to the last period of glaciation and global warming processes. Rates of subsidence along the Maine northern New England coast range from 0.06 in/yr (0.16 cm/yr) at Portland to 0.11 in/yr (0.34 cm/yr) at Eastport although in some regions it may approach 1.0 cm.[85] This rate may accelerate in coming years with the growing accumulation of carbon dioxide in the atmosphere. The insulating effects of CO^2 appear to be raising the average global temperature. Over the longterm, this increase could result in melting of the polar ice caps which in turn would raise the global sea level.

Sea level increases in the Bay of Fundy and the Gulf of Maine would enlarge the dimensions of the region's ocean basin. This lengthening could counteract the effects

Sea level increases in the Bay of Fundy and the Gulf of Maine would enlarge the dimensions of the region's ocean basin. This lengthening could counteract the effects of tidal power by pushing the Bay-Gulf system out of resonance. Examination of sediment cores from the upper Bay indicate that resonant effects have not been constant with time and suggest that with changes in water volume the system has gone in and out of resonance in the past.

Under average conditions today the tidal regime in this region exhibits a range between 60 and 140 percent. Some evidence for what a 10 percent increase may mean in ecological terms may be gained by examining the 115 km stretch of Maine coast between Bar Harbor and West Quoddy Head.[86] Here a 50 percent increase in tidal amplitude occurs.[87] Resident organisms seem resilient enought to withstand this variability.

Low stands in the recent past may give some indication of the effects of a lower low tidal range. If it is assumed that subsidence rates have been fairly constant over the last fifty years in the Eastport region, then in the 1930s low water was approximately 15 cm below where it is now. Projecting higher high tides using the same rate of subsidence suggests that tidal power effects will anticipate the high tide stand projected for fifty years. Projecting into the future, mean high tide will be roughly 15 cm higher in fifty areas, again assuming constant rates of submergence. This view tends to minimize the effects of a 10 percent tidal range shift. The effects of predicted eustatic sea level changes resulting from glacial melting, thus, would swamp both the effects of tidal power and local subsidence.

A contrary argument suggests that a tidal range increase would aggravate mean sea level increases. Following this reasoning, it has been calculated that a mean strip of land 2.5 meters (17 km^2 - 4,200 acres) wide would be flooded.[88] Another estimate places as much as 10,000 acres of coastal lowlands subject to inundation.[89] Both estimates derive from the view that coastal lowlands are presently subject to extensive erosion. It also assumes that the effects of mean sea level rise are roughly equivalent to those of tidal amplitude increases.[90] Confirmation of this contention requires empirical verification along the northeastern coast, particularly given the fundamental difference between tidal range and mean sea level.

Environmental Impacts of Other Tidal Power Facilities

Several areas could serve as analogs in the consideration of the effects of tidal power on the Bay of Fundy-Gulf of Maine environment. Two large-scale tidal power projects, the Rance River facility near St. Malo in France and the Kislaya Guba plant on the White Sea in the U.S.S.R., are now in operation although on a considerably smaller scale than what is projected for Fundy tidal power.[91] Another project has been proposed for the Severn River near Bristol, England. Feasibility studies have included a large scale ecosystem model which some suggest provides a good analog for the upper Fundy.[92] Further, the broad intertidal areas of the Netherlands and Germany along the North Sea encompass turbid, tidally-dominated environments similar to that of the upper Fundy.[93] Attempts to dam these areas indicate habitat and organism changes, resembling changes which might be expected in the upper Fundy.[94] Studies analyzing the feasibility of tidal power for the Passamaquoddy Bay and the environmental research preceeding construction of the Annapolis facility also suggest additional sources of information.[95] Finally, human-induced changes in the Fundy ecosystem, for example construction of the causeway across the Windsor mudflat, supply an immediate analog to Fundy tidal power.

Rance River Facility, France

The Rance plant, the first major hydro-electric facility to exploit the tides, began generating electricity in 1966.[96] It utilizes reversible blade turbines, generating 240 megawatts of electricity on the ebb and flow tides. The major concern at the time of plant construction was that high speed currents created by the dam would affect use of the nearby St. Malo harbor.[97] A lock was installed to minimize the effects of these currents on navigation. Because the Rance River carries little suspended particulate matter, little silting resulted from the facility. Loss of property reported around the headpond from the higher high tides has been minimal.[98] Electricite de France, the operator, however, has not conducted studies aimed at assessing the environmental impact of the facility. More distant effects such as along the southern coast of England have not been recorded, if observed, although the configuration of the English Channel and its connecting bodies of water suggest that increased resonance was not a factor.

Kislaya Guba, U.S.S.R.

Constructed in the late 1960s, this plant, located on the Kola Peninsula, uses a reverse blade turbine to generate 400 megawatts of electricity. Most available information on the facility deals with the engineering problems of construction in a harsh environment.[99] The facility is built across a narrow (40m) channel with fast moving currents. Tidal amplitude is only 2.5m. No information on environmental effects has been reported.

Severn River Barrage, U.K.

Like tidal power in the Bay and the Gulf, plans for a Severn facility date back to the early 1900s.[100] Currently an 800 megawatt ebb tide facility is under construction. The Bristol Channel-Severn estuary shares several similarities with the upper Fundy. It covers an area similar in size to that of the upper Fundy. It exhibits a wide tidal range (12m) and has large intertidal areas covered by turbid waters. In contrast, the Severn has a large freshwater input and little winter ice.

Modeling has traced the flow of organic carbon through the Bristol-Severn ecosystem.[101] It includes physical sub-routines, showing transfer rates of materials between sub-regions. The model accounts for river discharge, solar radiation, air temperature, wind stress, and shorebird and fish predation as forcing functions. Finally, it has accurately reflected major enviornmental perturbations experienced in the ecosystem since the model was constructed. Therefore, it is reasoned that the model provides a good predictor of Severn ecosystem response to barrage emplacement.[102]

Gordon and Longhurst have suggested that the Severn model with modification could be applied to the upper Fundy.[103] Holding the contrary view, Risk and Buckley have argued that the Fundy differs in several significant ways from the Severn system and that the de-dimensionalizing of the model necessary to allow it to operate would render the model useless in making predictions.[104] Risk and Buckley note that the Fundy exhibits higher tides, more well-mixed waters, and lower species diversities than does the Severn. They indicate that the Severn model is particularly weak in the areas of tidal modeling, sediment dynamics, and animal-sediment relationships - all important for understanding the upper Fundy. Finally, the Severn ecosystem shows high levels of primary production in contrast to the detritus-driven system of the Fundy. Thus, the energetics of the model may be inappropriate for the Fundy. Time for recovery after damming, for example, may be drastically different depending on how carbon is transferred in the two systems. In the Severn, the effects of damming may continue to favor higher primary productivity while in the Fundy damming would probably result in a shift from higher secondary productivity to higher primary productivity. No estimates exist to enlighten predictions of the time involved or the ecosystem implications of such a shift.

Passamaquoddy Studies

A two-basin tidal power facility was first suggested for the Passamaquoddy and Cobscook Bays in 1919, and this proposal stimulated investigation of potential environmental effects.[105] The report identified some changes, primarily disruption of migration routes of fish and loss of clam beds. More extensive studies were conducted in the 1950s when the Passamaquoddy proposal was revived.[106] These studies, conducted with the sponsorship of the International Joint Commission, concluded that impacts would be local and minor. The studies divided impacts into two categories: changes resulting from a shift in tidal amplitude and those stemming from a loss of access by fish to dammed-off waters.

Trites projected that the tides at the mouth of the Passamaquoddy Bay would increase by 31 cm (6 cm on the high tide and 6 cm on the low tide), although he hypothesized that the change would be too small to be detected.[107] Creation of headponds would alter the water column, eventually resulting in modification of the local climate. In the summer, reduced circulation in the headponds would result in water stratification. The warmer water would also contain about 50 percent less oxygen than would unstratified water. Sharper seasonal variation in the headponds would cause increased ice cover in the winter, causing harsher winter conditions.

Fisheries impacts included the destruction of the small herring fishery behind the barriers. The inward migration of cod, haddock, and pollock would cease.[108] Anadromous species, while losing some habitat, would benefit from warmer water temperatures as would larval lobster. Fishways could mitigate some fishery impacts caused by loss of access. Clam flats would be reduced and abundances would be further lowered. Clam populations, thought to be subject to cycles, were on a down phase and losses due to coastal flooding would exacerbate this decline. It was estimated it would take at least ten years for the stocks to rebuild if the dams were built.

Hachey and Huntsman (1933) and Hachey (1934), as reported by Scarratt and Dadswell, presented a dissenting view on the effects of tidal power on the Passamaquoddy ecosystem.[109] Hachey and Huntsman argued that restricting the flow of water into Cobscook and Passamaquoddy Bays would conserve energy otherwise dissipated and that this energy would increase tidal amplitudes elsewhere in the Fundy system. Scarratt and Dadswell also report that Huntsman believed that the high fisheries productivity in the mouth of the Bay of Fundy arose to a large degree because of outwelling and inwelling of water from the Passamaquoddy.[110] Altering this exchange would affect regional fisheries productivity. These views were not reflected in the 1935 report which concluded that the effects of damming would be minimal.[111] Huntsman, according to Scarratt and Dadswell, again raised these issues in the 1950s.[112]

The U.S. Army Corps of Engineers conducted a feasibility study for tidal power dams (18-970 megawatts) at a series of locations in Cobscook Bay, bordering the Passamaquoddy Bay.[113] The Corps also raised the issue of the effects on fisheries arising from loss of access, and a more fundamental issue of change in the trophic system resulting from an increase in tides. Their analysis, however, was directed at assessing economic feasiblity, and using that criterion, they concluded that tidal power was viable for Cobscook Bay.

Annapolis River, Nova Scotia

Interest in the environmental consequences of developing tidal power in the upper reaches of the Bay of Fundy generated curiousity about whether the environmental effects of the Annapolis River tidal project might function as an analog to the proposed Minas Basin project. Ecological research conducted in the Annapolis area suggests that the small-scale tidal project at Annapolis is a very poor paradigm for the large-scale Minas Basin project based on the following physical and biological differences.[114]

Besides the obvious differences in scale between the two areas, the two estuarine systems differ in their hydrographic characteristics. While the Annapolis river and headpond region are subject to variable conditions of water column stratification and while the Annapolis Basin is partially mixed, the greater turbulence of the water column throughout the Minas Basin system renders it completely mixed.[115] The extreme turbidity of the upper reaches also contrasts sharply with the lower turbidity of the

Annapolis estuary.[116] Because the original conditions of the Annapolis estuary are unknown, an unanswered question is the degree to which a causeway constructed in 1960 at the present tidal dam site has affected present conditions or how they may change when the tidal power dam begins operation.[117] However, Annapolis' biological communities were likely affected by construction of the causeway. Because the Minas Basin has never been previously dammed, the areas are not functioning under the same initial conditions so that the Annapolis project cannot be considered to serve as a control or reference to the Minas project. These physical differences have important biological consequences.

The Annapolis estuary supports a strong, diverse phytoplankton community that are at least moderately productive.[118] In the Minas Basin system, on the other hand, benthic species are the dominant primary producers owing to the extreme turbidity of the water column.[119] Basic differences in the relative contributions of primary producers yield differences in the structure, composition, and abundance of the biological communities in each area, with Annapolis being a more diverse, though not necessarily more productive, system. Species at the upper end of the food chain such as shad also apparently notice distinctions between the two estuaries since they use the Annapolis River for spawning and the Minas Basin for feeding.[120] Differing motivations for using each area, combined with the physical differences in estuary size, current strength, circulation pattern, and barrier size may also render Annapolis ineffective as an analog to the Minas project for predicting fish damage through turbine contact.

Many physical and biological differences between the two estuaries effectively remove Annapolis as a direct analog to the Minas project. However, studies in the Annapolis area have shown the strong influence that water column mixing has on both biological and physical/chemical features of the estuary. Although this is what would be expected in a dynamic environment like the Bay of Fundy, confirmation of this fact helps to form a conceptual model of how increases or decreases in water column turbulence resulting from barrier construction will be useful in refining predictions concerning the consequences of modifying the pattern and degree of mixing in Minas Basin.

Windsor Mudflat, Nova Scotia

Construction of a causeway at the southern bight of the Minas Basin resulted, over a ten year period, in exceptionally rapid accumulation of fine-grained sediment. The effects of this increase in sedimentation significantly altered the the biota. One view argues that Windsor provides an appropriate analog for what will occur when the Fundy tidal dams are emplaced. Another view, relying on modeling of sediment transport in the upper Fundy, suggests that in contrast to Cobequid Bay, the Windsor mudflat is a sediment sink.

Conclusion

None of the activities considered above provide a satisfactory analog for Fundy tidal power. Either information is missing (Rance and Kislaya Guba) or inappropriate for comparison. The role of organic carbon distinguishes Fundy from the Severn, with its primary productivity-driven communities, and from the lowlands of the Netherlands, which rely on the import of carbon. Passamaquoddy studies identify some impacts on the biota but generally conclude they are localized and small enough to be negligible. Tidal amplitude increases are the same as predicted for Fundy but, again, are concluded to be trivial. The Annapolis operation does provide a test of the utility of fishways, but again the environment of the Annapolis Basin is controlled by primary productivity and not the secondary productivity that plays an important role in the upper Fundy. The Windsor mudflat provides a useful analog to the degree it represents an end-member response for the upper Fundy. As mentioned though, none of the above studies considers environmental impacts beyond a relatively small geographic area. Thus, each study provides l imited insight into possible transboundary effects from the Fundy project, and this limitation suggests that attention should be focused on understanding the dynamics of the Bay of Fundy - Gulf of Maine system.

Specific Research Recommendations Concerning the Proposed Fundy Tidal Power Facility

The analysis presented in this chapter suggests that the transboundary, far-field, and indirect impact of a tidal facility in the upper Bay of Fundy would have little dramatic effect either on the New England coast or its marine resources. Damming the upper Bay would intensify existing forces in an ecosystem that has, through geological time, experienced considerable natural variation. The changes in tidal amplitude projected would not exceed what is experienced now in the Bay and the Gulf.

Nonetheless, assessing potential effects indicates important areas where our understanding of that natural variability is limited. Improved knowledge in some areas would contribute to our overall understanding. The areas include the effect that bottom topography and coastline configuration have on tidal energy dissipation, the effects of storms along the coastline, particularly in locations where, through human intervention, the coastline has been stabilized, and controls on the productivity of the marine ecosystem. Significant research is currently being conducted in all areas. This work could easily be expanded to address some of the more specific questions raised by the proposed Fundy facility. Outlined below are areas where additional information would be useful in resolving uncertainties about the effects of tidal power on the Bay of Fundy - Gulf of Maine region. Many of these are related to each other, but they are listed separately for emphasis.

Modeling (applicable to two and three dimensional models):

- studies of tidal energy dissipation, particularly in areas of rough bottom topography and indented coastlines;
- resolving depths in ares of rough bottom topography;
- improving accuracy of boundary conditions for both the open boundary with the North Atlantic and the coastal boundary;
- comparison of two and three dimension modeling efforts;
- linkage of "Greenberg-scale" modeling with coastal models to address localized or site-specific questions; and
- linkage of physical models with biological models of nutrient cycling, particularly in estuaries and at the interface of coastal and open Gulf of Maine waters.

Coastal Processes:

- site-specific studies of erosion, including definition of subtidal profiles and stability of inlets;
- radiocarbon dating of marsh peat cores to resolve questions about growth and development of coastal marshes and the effects of sea level and tidal amplitude changes;
- regional studies of estuarine areas to define nutrient cycling, productivity, and "overall health" of ecosystems (augmented by remote sensing); and
- intensified studies of physical and biological linkages between estuaries and more open waters.

Mixing and Circulation:

- effect of tidal mixing on current velocities, vertical mixing intensity, water temperature, and circulation patterns;
- relation of mixing processes to larval fish behavior, fish feeding, and fish migrations; and
- application of remote sensing techniques to defining mixing processes.

Notes

1. Charlier, R.H. 1982. Tidal Energy. Van Nostrand Reinhold. New York. pp. 234-274.

2. Bay of Fundy Tidal Power Review Board. 1977. Reassessment of Fundy Tidal Power.

3. Turk, T.R., M.J. Risk, R.W.M. Hirtle, and R.K. Yeo. 1980. Sedimentological and Biological Changes in the Windsor Mudflat, an Area of Induced Siltation. Canadian J. Fish. Aquat. Sci. Vol. 37. p. 1388.

4. Greenberg, D.A. 1980. Bay of Fundy Tidal Power - Mathematical Studies. Proc. Society for Industrial and Applied Mathematics. SIAM-SIMS Conference Ser. 7. pp. 51-82; Greenberg, D.A. 1979. A Numerical Model Investigation of Tidal Phenomena in the Bay of Fundy and Gulf of Maine. Marine Geodesy. Vol. 2. pp. 161-187.

5. *Ibid.*; Larsen, P.F. and J.A. Topinka. eds. 1984. Fundy Tidal Power Development: Preliminary Evaluation of its Environmental Consequences to Maine. A Report to the Maine State Planning Office; Spiller, J. Ed. Regional Scientific Needs in Fundy Tidal Power Development - Report of a Workshop. Working Paper 3. International Risks of Coastal Energy Development in the Gulf of Maine - Bay of Fundy Paper. Marine Law Institute. University of Southern Maine.

6. Bumpus, D.F. 1974. A Description of the Circulation on the Continental Shelf of the East Coast of the United States. Progress in Oceanography. Vol. 2. pp. 11-157; Butman, B. et al. 1982. Recent Observations of the Mean Circulation on Georges Bank. Jour. Phys. Oceanography. Vol. 12; Hachey, H.B. and W.B. Bailey. 1957. The General Hydrography of the Waters of the Bay of Fundy. Jour. Fish. Res. Bd. Canada. Manuscript Report of the Biological Stations No. 633.

7. *Ibid.*, Bumpus.

8. *Ibid.*

9. Iles, T.D. and M. Sinclair. 1982. Atlantic Herring: Stock Discreteness and Abundances. Science. Vol. 215.

10. Sutcliffe, W.H., R.H. Loucks, and K.F. Drinkwater. 1976. Coastal Circulation and Physical Oceanography of the Scotian Shelf and the Gulf of Maine. Jour. Fish. Res. Bd. Canada. Vol. 33.

11. Kulka, D.W., S. Corey, and T.D. Iles. 1982. Community Structure and Biomass of Euphausiids in the Bay of Fundy. Can. Jour. Fish. Aquat. Sci. Vol. 39.

12. Greenberg, *supra* note 4.

13. Dadswell, M.J., G.D. Melvin, P.J. Williams, and G.S. Brown. 1984. Possible Impact of Large-Scale Tidal Power Developments in the Upper Bay of Fundy on Certain Migratory Fish Stocks of the Northwest Atlantic. In D.C. Gordon, Jr., and M.J. Dadswell. Eds. *Update on the Marine Environmental Consequences of Tidal Power Development in the Upper Reaches of the Bay of Fundy*. Canadian Tech. Rept. of Fish. and Aquat. Sciences No. 1256.

14. Cetacean and Turtle Assessment Program. 1982. A Characterization of Marine Mammals and Turtles in the Mid-and North-Atlantic Areas of the U.S. Continental Shelf. U.S. Dept. of Interior, Minerals Management Service Contract No. AA551-CT8-48.

15. Brown, W.S. 1984. A Comparison of Georges Bank, Gulf of Maine, and New England Shelf Tidal Dynamics. Jour. Phys. Ocean. Vol. 14; Greenberg, D.A. 1983. Modeling the Mean Barotropic Circulation in the Bay of Fundy. Jour. Phys. Ocean. 13; Ku, L., D.A. Greenberg, C.J.R. Garrett, and F.W. Dobson. 1985. Nodal Modulation of the Lunar Semidiurnal Tide in the Bay of Fundy and Gulf of Maine. Science. Vol. 230.

16. Greenberg, *supra* note 4.

17. Bunton, G.W. 1966. *The Tides*. British Museum Press. Hawaii. p. 7.

18. Pond, S. and G.L. Pickard. 1978. *Introductory Dynamic Oceanography*. Pergamon Press. New York. p. 200.

19. *Ibid.*, p. 201.

20. *Ibid.*

21. *Ibid.*, p. 202.

22. *Ibid.*, p. 203.

23. *Ibid.*

24. *Ibid.*

25. Hachey and Bailey, *supra* note 6.

26. Pond and Pickard, *supra* note 18 at 205.

27. Amos, C.L. and B.F.N. Long. 1980. The Sedimentary Character of the Minas Basin, Bay of Fundy. In S.B. McCann. Eds. *The Coastline of Canada: Littoral Processes and Shore Morphology*. Geol. Survey Canada. Paper 80-10; Greenberg, *supra* note 4.

28. Garrett, C.J.R. 1974. Tidal Resonance in the Bay of Fundy and Gulf of Maine. Nature. Vol. 238. pp. 549-556; Normal Modes of the Bay of Fundy and Gulf of Maine. Can. Jour. Earth Science. Vol. II.

29. *Greenberg*, *supra* note 4.

30. Duff, G.F.D. 1970. Tidal Resonances and Tidal Barriers in the Bay of Fundy System. Jour. Fish. Res. Bd. Canada. Vol. 27. p. 1701; Garrett, *supra* note 28; Greenberg, *supra* note 4; Parkinson, R.E. 1972. Mathematical Model of the Tidal Regime in the Bay of Fundy. In O.K. Gashus and T.S. Gray. Eds. *Tidal Power*. Plenum. New York. p. 131; Proudman, J. 1953. *Dynamical Oceanography*. Metheun and Co., Ltd. London. p. 409; Rao D.B. 1968. Natural Oscillations in the Bay of Fundy. Jour. Mar. Res. Vol. 25 p. 1907; Yuen K.B. 1969. Effect of Tidal Barriers on the M Tide in the Bay of Fundy. Jour. Fish. Res. Bd. Canada. Vol. 26. p. 2477.

31. Proudman, *supra* note 30.

32. Rao, *supra* note 30.

33. Redfield, A.C. 1953. Interference Phenomena in the Tides of the Woods Hole Region. Jour. Mar. Res. Vol. 12 pp. 121-140.

34. Ippen, A.T. and D.R.F. Harleman 1966. *Estuary and Coastline Hydrodynamics*. McGraw Hill. New York.

35. Yuen, *supra* note 30.

36. Duff, *supra* note 30.

37. Garrett, *supra* note 28.

38. *Ibid.*

39. *Ibid.*

40. Greenberg, *supra* note 4.

41. *Ibid.*

42. Tatsusaburo, I. and M.L. Spaulding. 1984. A Model of Tidally Induced Residual Circulation in the Gulf of Maine and Georges Bank. Jour. Phys. Ocean. Vol. 14. p. 1121.

43. Greenberg, D.A. 1977. Mathematical Studies of Tidal Behavior in the Bay of Fundy. Ph.D. Thesis. Univ. of Liverpool. England. p. 74.

44. Heaps, N.S. 1972. Tidal Effects Due to Water Power Generation in the Bristol Channel. In O.K. Gashus and T.S. Gray, Ed. *Tidal Power*. Plenum Press. New York. pp. 435-455.

45. Dadswell et al., *supra* note 13 at 581; Scarratt, D.J. and M.J. Dadswell. 1982. The Effect of Tidal Power Generating Systems on Local and Migrant Fish Populations. Paper Prepared for New Approaches to Tidal Power. Conference held Halifax, Nova Scotia. June 1-4, 1982. p. 1.

46. *Ibid.*

47. *Ibid.*, Dadswell et al. pp. 589, 594.

48. *Ibid.*, p. 594.

49. *Ibid.*, p. 596.

50. *Ibid.*, p. 597.

51. Melville and M.J. Dadswell. 1984. Possible Impact of Tidal Power Development on the Spawning Population of American Shad (*Alosa sapidissima*) in the Annapolis River, N.S. In D.C. Gordon, Jr., and M.J. Dadswell. Eds. *Update on the Marine Environmental Consequences of Tidal Power Development in the Upper Reaches of the Bay of Fundy*. Canadian Tech. Report Fish. and Aquat. Sci. No. 1256. p. 543.

52. Greenberg, D.A. and C.L. Amos. 1983. Suspended Sediment Transport and Deposition Modeling in the Bay of Fundy, Nova Scotia - A Region of Potential Tidal Power Development. Canadian Jour. Fish. Aquat. Sci. Vol. 40. pp. 27, 28.

53. Smith, P.C. and P.W. Hicklin. 1984. Potential Impacts of Tidal Power Developments on Bird Populations in the Bay of Fundy. In D.C. Gordon, Jr. and M.J. Dadswell. Eds. *Update on the Marine Environmental Consequences of Tidal Power Development in the Upper Reaches of the Bay of Fundy*. Canadian Tech. Report of Fish. and Aquat. Sciences No. 1256.

54. Personal communication. Dr. David Scarrett. St. Andrews Biological Station. Canadian Department of Fisheries and Oceans (November 1983).

55. Pond and Pickard, *supra* note 18 at 209.

56. *Ibid.*

57. Miller, A.R. 1958. The Effects of Winds on Water Levels on the New England Coast. Limn. and Oceanogr. Vol. 3. p. 5.

58. Tiner, R. U.S. Fish and Wildlife Service, Newton Corner, Ma. Internal Memo to D. Riley. Shoreline Damages Related to Bay of Fundy Tidal Power Project (January 29, 1982).

59. Miller, *supra* note 57 at 10.

60. *Ibid.*, p. 13.

61. U.S. Army Corps of Engineers. 1980. Tidal Flood Profiles. New England Coastline. Hydraulics and Water Quality Division. Paltes C24- C29.

62. Tiner, *supra* note at 58.

63. Personal communication L.K. Fink. Professor of Geology. University of Maine (November 1983).

64. Larsen and Topinka, *supra* note 5 at 62.

65. Nixon, S.W. 1982. The Ecology of New England High Salt Marshes: A Community Profile. U.S. Fish and Wildlife Service. FWS/OBS-81/55.

66. See, for example, Larsen and Topinka, *supra* note 5 at 56; Spiller, *supra* note 5 at 7-9.

67. Iles and Sinclair, *supra* note at 9.

68. *Ibid.*

69. *Ibid.*

70. *Ibid.*

71. Garrett, C.J.R., J.R. Keeley, and D.A. Greenberg. 1978. Tidal Mixing Versus Thermal Stratification in the Bay of Fundy and Gulf of Maine. Atmosphere-Ocean. Vol. 16.

72. Simpson, J.H. and J.R. Hunter. 1974. Fronts in the Irish Sea. Nature Vol. 250.

73. Garrett et al. *supra* note 71 at 421.

74. Dickie, L.M. 1955. Fluctuations in Abundance of the Giant Scallop, *Placopecten magellanicus*, in the Digby area of the Bay of Fundy. Jour. Fish. Res. Bd. Canada. Vol. 12. p. 843; Caddy, J.F. 1979. Long-term Trends and Evidence for Production Cycles in the Bay of Fundy Scallop Fishery. Rapp. P. -V. Reun. Cons. It. Explor. Mer. Vol. 175. pp. 97-108.

75. *Ibid.*, Dickie, p. 874.

76. *Ibid.*

77. New England Fishery Management Council. 1982. Environmental Impact Statement for the Atlantic Sea Scallop (*Placopecten magellanicus*) Fishery Management Plan.

78. Dickie, *supra* note 74 at 852.

79. MacKay, A.A., Bosien, R.K., and P. Leslie. 1979. Bay of Fundy Resource Inventory. Vol. 4. The Grand Manan Archipelago. New Brunswick Department of Fisheries NB78-1A.

80. Caddy, *supra* note 74 at 106.

81. Caddy, *supra* note 74; Dickie, *supra* note 74 at 371.

82. Doggett, L. and S. Sykes. 1980. Commercially Important Invertebrate. In U.S. Fish and Wildlife Service. *An Ecological Characterization of Coastal Maine*. Vol. 3. p. 10.

83. White, A.W. and C.M. Lewis. 1982. Resting Cysts of the Toxic Red Tide Dinoflagellate, *Gonyaulax excavata*, in the Bay of Fundy Sediments. Canadian Jour. Fish. Aquat. Sci. Vol. 39. p. 1190.

84. *Ibid.* p. 1193.

85. Grant, D.R. 1968. Recent Submergence in Nova Scotia and Prince Edward Island, Canada. Geol. Surv. Canada Paper No. 68. p. 162-164; Hicks, S.D. 1978. An Average Geopotential Sea Level Series for the United States. Shore and Beach. pp. 20-23.

86. Desplanque, C. 1984. Comments on Larsen and Topinka, *supra* note 5. pp. 1-7.

87. Review of tide ranges in Eldredge. 1984. Tide Tables.

88. Larsen and Topinka, *supra* note 5 at 58.

89. *Ibid.*

90. *Ibid.*, p 57.

91. Charlier, *supra* note 1.

92. Gordon, D.C., Jr. and A.R. Longhurst. 1982. The Environmental Aspects of a Tidal Power Project in the Upper Reaches of the Bay of Fundy. Mar. Poll. Bull. Vol. 10, pp. 38-45.

93. Gordon, D.C., Jr. and J.W. Baretta. 1982. A Preliminary Comparison of Two Turbid Coastal Ecosystems: the Dollard (Netherlands-FRG) and the Cumberland Basin (Canada). Hydrobio. Bull. Vol. 19.

94. *Ibid.*

95. International Passamaquoddy Engineering Board. 1959. Investigation of the International Passamaquoddy Tidal Power Project. Report to the International Joint Commission. Appendix III. Observation and Prediction of Tides; MARTEC. 1980. Annapolis Tidal Power Proejct: Environmental Impact Statement.

96. Charlier, *supra* note 1.

97. *Ibid.*

98. *Ibid.*

99. *Ibid.*

100. *Ibid.*

101. Longhurst, A.R. 1978. Ecological Models in Estuarine Management. Oceanology Int. Vo. 78. pp. 91-4; Radford, P.J. and R.H. Uncles. 1980. Ecosystem models and the Prediction of Ecological Effects. In T.S. Shaw. Ed. *Environmental Appraisal of a Tidal Power Station: with Particular Reference to the Severn Barrage*. Pitman Publishing, Ltd. London. pp. 109-127.

102. Gordon and Longhurst, *supra* note 92.

103. *Ibid.*

104. Risk, M.J. and D.E. Buckley. 1979. Viewpoint on a Viewpoint. Mar. Poll. Bull. Vol. 10. pp. 219-222.

105. North American Council on Fisheries Investigations. 1935. The Effects of Dams in the Passamaquoddy Bay. Proc. North American Council on Fisheries Investigations. 1931-1933.

106. Hart, J.L. and D.L. McKernan. 1960. International Passamaquoddy Fisheries Board Fisheries Investigation. Introductory Accout. Jour. Fish. Res. Bd. Canada Vol. 17. pp. 127-131; International Passamaquoddy Engineering Board, *supra* note 95; Trites, R.W. 1961. Probable Effects of the Proposed Passamaquoddy Tidal Power Project on Oceanographic Conditions. Jour. Fish. Res. Bd. Canada Vol. 18. pp. 163-201.

107. *Ibid.*, Trites.

108. Hart and McKernan, *supra* note 106.

109. Hachey, H.B. 1934. The Probable Effect of Tidal Power Development on the Bay of Fundy Tides. Jour. Franklin Institute. Vol. 217. pp. 747-756; Hachey, H.B. and A.G. Huntsman. 1933. The Cooper Power Project and the Probable Effects on the Bay of Fundy Tides. Jour. Fish. Res. Bd. Canada Manuscript Report No. 62. Both cited in Scarratt and Dadswell, *supra* note 45.

110. Huntsman, cited in *Ibid*, Scarratt and Dadswell.

111. North American Council on Fisheries Investigations, *supra* note 105.

112. Scarratt and Dadswell, *supra* note 45.

113. U.S. Army Corps of Engineers. 1980. Investigations of Tidal Power - Cobscook Bay, Maine. New England Division.

114. Daborn, G.R., A.M. Redden, and R.S. Gregory. 1982. Ecological Studies of the Annapolis Estuary. The Acadia University Institute No. 29.

115. *Ibid.*; Amos and Long, *supra* note 27.

116. Amos, C.L. and T.T. Alfoldi. 1979. The Determination of Suspended Sediment Concentration in a Macrotidal System Using Landsat Data. Jour. Sed. Petrology. Vol. 49.

117. Daborn et al., *supra* note 114.

118. *Ibid.*

119. Prouse, N.J., D.C. Gordon, Jr., B.T. Hargrave, C.J. Bird, J. McLachlan, J.S.S. Lakshminarayana, J. Sita Devi, and M.L.H. Thomas. 1984. Primary Production: Organic Matter Supply to Ecosystems in the Bay of Fundy. In D.C. Gordon, Jr. and M.J. Dadswell. *Update on the Marine Environmental Consequences of Tidal Power Development in the Upper Reaches of the Bay of Fundy.* Canadian Tech. Report Fish. and Aquat. Sciences No. 1256.

120. Melville and Dadswell, *supra* note 51.

Chapter 3

Legal Decisionmaking Frameworks for Fundy Tidal Power

David VanderZwaag and Karen A. Massey

Introduction

This chapter provides an overview of the Canadian and U.S. decisionmaking processes potentially applicable to Fundy tidal power and highlights the legal and administrative uncertainties inherent in federal review mechanisms, state and provincial permit procedures, and judicial proceedings. The first section examines the U.S. framework of state and federal laws which would impose technological and environmental review requirements on a domestic hydro-electric project or a Canadian Fundy tidal power project. The chapter next reviews the Canadian legal framework for regulating tidal power development and focuses on the regime most likely to apply to the Fundy Tidal Power proposals. The final section compares the American and Canadian legal systems for controlling hydro-electric projects and concludes by analyzing which system is better equipped to address concerns over long-range, indirect environmental effects. Chapter 4 then examines this question in considerable detail, focusing on specific instances where the two systems were called upon to consider indirect and transboundary environmental effects.

The United States Decisionmaking Framework

U.S. Regulation of Domestic Tidal Power Development[1]

A tidal power project is in many ways similar to a conventional hydro-electric project, and the development of such a project in the United States would be subject to the same environmental constraints and permitting procedures. The rapid rise in oil prices in the past decade prompted a renewed interest in hydro-electric power that spurred the U.S. Congress to pass legislation encouraging hydro-power development.[2] The result has been considerable activity in the examination of hydro-power resources for potential development. This activity has in turn produced documents and articles outlining the regulatory procedures for obtaining the necessary permits to operate a hydro-electric facility.[3] These regulatory analyses are transferrable to a tidal power development proposal with some modifications because of the development's coastal rather than inland location. The relevant statutes and regulatory agencies are discussed below in the following order: 1) jurisdictional legislation; 2) regulatory statutes and agencies; 3) environmental overview under the National Environmental Policy Act (NEPA) and Executive Order 12,114; 4) related special resources protection legislation (which in many cases predated NEPA, but which continues in force and supplements NEPA); and 5) the Coastal Zone Management Act and other state controls.

1. *Jurisdictional Legislation: The Submerged Lands Act*

The Submerged Lands Act (SLA),[4] enacted in 1953, in general confirmed to the states (or to those persons "entitled thereto under the law of the respective states') *ownership* of the lands beneath navigable waters within the boundaries of those states and seaward from the coastline three geographical miles into the Atlantic and Pacific Oceans and three marine leagues into the Gulf of Mexico. However, one subsection of that act[5] provides:

> Nothing in this Chapter shall affect the use, development, improvement or control by or under the constitutional authority of the United States of said lands and waters for the purposes of navigation or flood control *or the production of power*, or be construed as the release or relinquishment of any rights of the United States arising under the constitutional authority of Congress to regulate or improve navigation, or to provide for flood control, or the *production of power*.

(Emphasis added.)

Thus the SLA would not provide a basis for state regulation of tidal power development activities sanctioned by the federal government, but would seem to confirm the control of such activities as in the federal domain. It would, however, be necessary for a federal or private tidal power developer to obtain a sale or lease of submerged lands from the state or person holding title under state law[6] or to take the land by eminent domain, paying just compensation therefore.[7]

2. *Regulatory Legislation*

The Federal Power Act of 1920 (FPA),[8] administered by the Federal Energy Regulatory Commission (FERC), provides the primary system for the construction and operation of dams that would cover a tidal power dam. It authorizes the Commission

> [t]o issue licenses . . . for the purpose of constructing, operating, and maintaining dams, water conduits, reservoirs, power houses, transmission lines, or other project works necessary or convenient . . . for the development, transmission, and utilization of power across, along, from, or in any streams or other bodies of water over which Congress has jurisdiction . . . *Provided* . . . [t]hat no license affecting the navigable capacity of any navigable waters of the United States shall be issued until the plans of the dam or other structure affecting navigation have been approved by the Chief of Engineers and the Secretary of the Army. Whenever the contemplated improvement is, in the judgment of the Commission, desirable and justified in the public interest for the purpose of improving or developing a waterway . . . for the use or benefit of interstate or foreign commerce, a finding to that effect shall be made by the Commission.[9]

The FPA further provides that a licensed project:

> shall be such as in the judgment of the Commission will be best adapted to a comprehensive plan for improving or developing a waterway or waterways for the use or benefit of interstate or foreign commerce, for the improvement and utilization of waterpower development, and for other beneficial public uses, including recreational purposes.[10]

Prior to the passage of the National Environmental Policy Act (NEPA), (discussed below) the "public interest" and "best adapted" language of these provisions already had been interpreted to require a general "public interest review" of a project before licensing. That review should encompass an inquiry into such questions as the preservation of fisheries for commercial and recreational purposes and the protection of wildlife and their habitat.[11] However, according to one commentator, the reports that serve as FERC comprehensive plans:

> usually are devoted to insuring that one hydro-electric project will not interfere with the generating potential of those already in existence or planned and that the maximum power potential of the river is developed. One looks in vain for explicit description, much less balancing, of all three statutory interests [interstate or foreign commerce, water power development, and other beneficial public uses].[12]

The Federal Power Act is arguably, by court interpretation, the only regulatory authority, federal[13] or state,[14] applicable to hydro-power development, presumably including tidal power. However, because there has been no clear declaration by the United States Supreme Court that all other potentially applicable federal controls are preempted by the Federal Power Act, these federal controls will be discussed. There will also be a brief consideration of state controls, for even there Supreme Court interpretations may shift after almost forty years.[15]

Section 9 of the Rivers and Harbors Act of 1899[16] requires both Congressional authorization and plan approval by the Army Corps of Engineers for the construction of any bridge, dam, dike, or causeway over or in navigable waters of the United States. Section 10 of the Rivers and Harbors Act,[17] which is the major regulatory control for harbor and channel dredging projects, requires Army Corps authorization "to excavate or fill, or in any manner to alter or modify the course, location, condition, or capacity of" navigable waters. The Corps approvals required are exercised by imposition of conditions via the FERC licensing process rather than through separate Corps permits.[18]

Section 404 of the Clean Water Act regulates the discharge of dredged or fill material into waters of the United States.[19] As the construction of a new tidal power dam would require the placement of "fill,"[20] it would require a § 404 permit.[21]

One possibly applicable exception is that the section excludes from permitting requirements:

> the discharge of dredged or fill material as part of the construction of a Federal project specifically authorized by Congress ... if information on the effects of such discharge ... is

> included in an environmental impact statement ... submitted to Congress before the actual discharge of dredged or fill material in connection with the construction of such project and prior to either authorization of such project or an appropriation of funds for such construction.[22]

If the Army Corps of Engineers, therefore, were to develop the project, it would not have to obtain a permit from itself, but would have to prepare an EIS.

Although the § 404 regulations no longer require that an applicant receive all other applicable federal, state, and local approvals before a Corps permit can be granted, the denial of another permit during processing of the § 404 permit will result in denial of the dredge and fill permit without prejudice to reinstate processing should the denied approval later be received.[23] Failure to receive State Water Quality Certification required under § 401 of the Clean Water Act[24] will result in permit denial as will failure to get state concurrence in the certification of consistency with the Coastal Zone Management Act.[25]

Section 404 requires the Corps to perform a public interest review--a balancing of the adverse impacts of the project against the benefits. Evaluation of adverse environmental impacts is on the basis of regulatory guidelines.[26] These guidelines are developed by the Environmental Protection Agency, "in conjunction with" the Corps.[27] The guidelines identify the protection of aquatic ecosystems, particularly wetlands, from damaging dredge and fill discharges as a principal aim.[28] In addition, if EPA objects to the issuance of a Corps permit on environmental grounds, it can force the resolution of permit issuance upward through the administrative hierarchy of the two agencies. In the case of the construction of a new tidal power dam, however, the issuance of the permit would have to be decided at the top administrative level, the Chief of Engineers.[29] The Chief can override EPA objections if the project is necessary for navigation. EPA, however, has the ultimate authority: Section 404 authorizes the EPA Administrator to prohibit or restrict the use of any defined area for dredge or fill disposal "whenever he determines, after notice and opportunity for public hearings, that the discharge of such materials into such area will have an unacceptable adverse effect on municipal water supplies, shellfish beds and fishery areas..., wildlife, or recreational areas."[30] However, this authority has seldom been invoked.[31]

In practice, the Corps public interest review may, at least in part, be subsumed in the FERC licensing proceedings. Because of its broad authority over hydro-electric projects FERC would almost certainly be the lead agency for preparation of an Environmental Impact Statement.

3. *Environmental Overview Authority*

In 1970, the United States Congress passed NEPA,[32] requiring all federal agencies taking any actions, including licensing, that would have a "significant effect on the quality of the human environment"[33] to perform an overall environmental review and assessment not unlike that already required under the Federal Power Act. The Council on Environmental Quality (CEQ) has formulated general regulations to guide the agencies in the implementation of NEPA and the preparation of an Environmental Impact Statement (EIS).[34] Although there are scientific information requirements for all environmental permits, the intent of the NEPA regulations is that scientific data needs for all permits be addressed in the EIS, when one is required. The EIS is intended to be the documentary basis for decisionmaking. A recent CEQ document providing guidance to the agencies on the implementation of CEQ's NEPA regulations describes the process of "scoping" (determining the scope of the EIS):

> The scoping process should identify the public and agency concerns; clearly define the environmental issues and alternatives to be examined in the EIS including the elimination of nonsignificant issues; identify related issues which originate from separate legislation, regulation, or Executive Order (e.g., historic preservation or endangered species concerns); and identify state and local agency requirements which must be addressed.[35]

Most of the CEQ regulatory requirements with respect to preparation of EIS's are qualitative, however, and do not provide clear standards for judging whether the scientific information available is adequate to the task at hand or has been used in an unbiased manner in the decisionmaking process. The regulations rely primarily on procedural controls to insure that the science is adequate to the task and is incorporated into decisionmaking. The controls include: timing, i.e., that the document is prepared before a decision is made; peer review through interagency consultation with agency staff with expertise in, and responsibility for, specific resources, e.g., fish and wildlife; public review, which presumably will provide a diversity of perspectives to make certain that all important issues are raised; and consideration of alternatives to the proposed action.[36] The regulations identify this last consideration as "the heart of the environmental impact statement."[37]

The regulations do provide a method for dealing with incomplete or unavailable information.

> (a) If the information relevant to adverse impacts is essential to a reasoned choice among alternatives and is not known and the overall costs of obtaining it are not exorbitant, the agency shall include the information in the environmental impact statement.
>
> (b) If (1) the information relevant to adverse impacts is essential to a reasoned choice among alternatives and is not known and the overall costs of obtaining it are exorbitant or (2) the information relevant to adverse impacts is important to the decision and the means to obtain it are not known ... the agency shall weigh the risk and severity of possible adverse impacts were the action to proceed in the face of uncertainty. If the agency proceeds, it shall include a worst case analysis and an indication of the probability or improbability of its occurence.[38]

FERC has promulgated regulations incorporating the NEPA and CEQ requirements for preparation of an Environmental Impact Statement (EIS) into its licensing procedures

> [f]or a major hydro-electric project (capacity greater than 2000 hp or 1.5 MW) that is completely unconstructed, or that includes a proposed change in the existing state of project

works or project operation which would result in a significant change in the normal maximum surface area or normal maximum surface elevation of an existing impoundment or in a significant environmental impact....[39]

For our purposes this definition can be assumed to cover any tidal power development. Certainly development of tidal power on the scale now proposed by Nova Scotia, or on the scale already developed at Annapolis Royal, would require preparation of a detailed Environmental Impact Statement. A worst case analysis would no doubt be required for a major tidal power development were it to occur in the United States jurisdiction, because of the uncertainties surrounding use of a relatively new type of project on such a large scale.[40]

NEPA only mandates the preparation of a document. While agencies can be and have been sued under NEPA for failure to prepare, or to properly prepare, an impact statement, thus delaying projects, NEPA itself cannot be invoked to prevent the licensing or construction of any project, even if significant environmental impact from the activity is evident from the EIS. In short it is not considered to be a regulatory, but merely an informational and advisory statute.

While such an impact statement would have to deal fully with environmental impacts within United States jurisdiction, the necessity to treat and extent of treatment of impacts abroad, e.g., in Canada, under NEPA is uncertain as will be further discussed in the following sections.

On January 4, 1979, President Carter signed Executive Order 12,114, "Environmental Effects Abroad of Major Federal Actions."[41] The Order was intended to clarify the scope of federal agencies' responsibilities to evaluate the environmental effects of their actions abroad. The issue had been one of considerable controversy ever since the passage of NEPA, producing a series of administrative decisions and court cases[42] and much related law review commentary.[43] The controversy came to a head in late 1977 and early 1978 when the Council on Environmental Quality circulated among the agencies its proposed regulations on preparation of Environmental Impact Statements for United States actions abroad.[44] The State Department strongly objected to CEQ's position on the extent of the extraterritorial applicability of NEPA. The final arbiter then became the President via Executive Order 12,114. The Order itself seems to have exacerbated, rather than resolved, the confusion and controversy surrounding the question of NEPA's applicability abroad, as evidenced by the extensive legal commentary that it has provoked.[45]

The Order is fairly short, but comprehensive. First, it states that it is based on "independent authority,"[46] but in general furthers the purposes of NEPA, the Marine Protection Research and Sanctuaries Act (MPRSA),[47] and the Deepwater Port Act,[48] "consistent with the foreign policy and national security policy of the United States." Significantly, this opening section, entitled "Purpose and Scope", further states that the Order

> represents the United States government's exclusive and complete determination of the procedural and other actions to be taken by Federal agencies to further the purpose of the National Environmental Policy Act ... with respect to the environment outside the United States, its territories and possessions.[49]

It then provides for the development of agency procedures for four types of "major Federal actions": 1) those actions "significantly affecting the environment of the global commons outside the jurisdiction of any nation"; 2) actions significantly affecting the environment of a foreign nation not participating with the United States and not otherwise involved in the action; 3) actions significantly affecting the environment of a foreign nation that provide to that nation certain products or projects strictly regulated in the United States because of toxicity or radioactivity; and 4) actions outside the United States' jurisdiction affecting natural resources of global importance specifically designated for protection pursuant to the order by the President--or by the Secretary of State if in accord with a binding international agreement.[50] It also provides for various types of environmental studies including environmental impact statements, bilateral or multilateral studies, and "concise reviews of the environmental issues involved, including environmental assessments, summary environmental analyses and other appropriate documents."[51] A full-blown environmental impact statement is required *only* for those actions significantly affecting the global commons. And even in those cases the impact statement would differ from a traditional domestic EIS in that social and economic impacts need not be considered, only impacts on the physical environment.[52] There follows a lengthy list of "Exemptions and Considerations"[53] potentially, unless very strictly interpreted, broad enough to swallow the requirement to prepare (or, if prepared, to release) any environmental assessment of effects abroad of United States actions.

Section 3 provides in significant part first that "nothing in this Order shall be construed to create a cause of action"[54] and further that

> [i]f a major Federal action having effects on the environment of the United States or the global commons requires preparation of an environmental impact statement, and if the action also has effects on the environment of a foreign nation, an *environmental impact statement* need not be prepared with respect to the effects on the environment of the foreign nation.[55]

The meaning of this final provision is ambiguous in the context of the entire Order. The Order never requires environmental impact statements to be prepared for effects on a foreign state; thus the quoted section appears superfluous. The subsection may, however, have been deemed necessary to clarify that where an EIS is otherwise required the existence of effects on other jurisdictions is not subsumed within that requirement, but will continue to be governed by the requirements of the Order directly applicable (although the language appears to speak of a separate environmental impact statement). In any event, as the Order clearly differentiates between Environmental Impact Statements and other types of environmental assessments, this provision would not affect the requirement for preparing these studies and/or assessments in accordance with Executive Order 12,114.[56]

Thus, under the Executive Order, only some sort of environmental assessment less than an EIS would be required with respect to the environmental effects in Canada of a tidal power development in Maine. In addition, if the project were a joint venture with the Canadian government or an agency thereof, the Order would require no assessment of the impacts on Canada.[57] Finally the Order would appear to preclude any action by Canadians to enforce even that requirement. The scenario that occurred in *Wilderness Society v. Morton*,[58] of a Canadian environmental group suing to enforce NEPA, is arguably precluded by the Order.

The question remains whether the President can, in fact, by Executive Order, essentially override the Congressional mandates of NEPA, assuming that statute has extraterritorial impact. At least one commentator argues that "the courts will likely factor the Executive Order lightly, if at all, when next faced with the question of the international reach of NEPA" because it strays too far from the statutes, although it may serve as a guide on the extent to which environmental analysis is required when there are conflicts with foreign policy.[59] Another commentator suggests a low likelihood of success in a suit brought under NEPA now to challenge effects of an action abroad, noting the traditional deference of the court to the President in foreign policy matters.[60] These sorts of separation of power issues are the type of political question courts prefer to avoid.

Absent a treaty varying the terms of Executive Order 12,114, the extent of environmental assessment of tidal power impacts under the Order would be largely in the discretion of the agencies preparing the assessment. In the case of tidal power, however, if the impacts on the Canadian environment would not be significantly different from those on the Maine coast and waters, the Canadians should at least be as fully apprised of those effects as the Americans, as an EIS would be prepared for effects of the project within the United States, and those effects would clearly be a consideration in the determination to construct the project.

4. Special Resource Protection Legislation

The Fish and Wildlife Coordination Act of 1956 (FWCA)[61] requires applicants to agencies for permits for activities that involve the control or modification of any body of water to consult first with state and federal fish and wildlife agencies "with a view to the conservation of wildlife resources by preventing loss of and damage to such resources as well as providing for the development and improvement thereof in connection with such water resource development."[62] Permitting agencies such as FERC, the Army Corps of Engineers, and EPA must give full consideration to the reports, comments, and recommendations of these fish and wildlife agencies in the permitting process.[63]

The FWCA is to some extent duplicative of NEPA, which it predated, in requiring consultation with federal and state agencies with jurisdiction over particular environmental impacts and the consideration of impacts on fish and wildlife resources.[64] However, the FWCA requires consideration not just of mitigation of impacts in connection with water resources development, but also of *enhancement* of these resources in conjunction with the development. The act states that "the project plan *shall include* such justifiable means and measures for wildlife purposes as the reporting agency[65] finds should be adopted to obtain maximum overall project benefits."[66]

Because of the explicit authority of the FWCA, recommendations of the federal and state fish and wildlife agencies are taken very seriously. Though development may be approved that they object to, their recommended mitigation measures are usually included by FERC as project conditions, although such inclusions are mandatory only in limited cases.[67]

The Endangered Species Act[68] and the Marine Mammal Protection Act[69] were enacted in 1973 and 1972 respectively to protect the world's endangered and threatened species. The latter statute gives special attention to marine mammals, many of which are endangered or threatened species and others of which are "depleted" by current practices that emphasize short-term economic gain. The statute cites marine mammals as "resources of great international significance, esthetic and recreational as well as economic."[70]

Because of the high tides in eastern Maine, this area, particularly Cobscook and Passamaquoddy Bays, is the most likely for a tidal development project in the United States. As the Pittston oil refinery cases[71] have shown, the Endangered Species Act is likely to come into play in this area because it serves as a habitat for the bald eagle and for certain endangered species of whales. Section 7 of the Act as amended, provides *inter alia* that

> Each federal agency shall ... insure that any action authorized, funded or carried out by such agency ... is not likely to jeopardize the continued existence of any endangered species or threatened species or result in the adverse modification of [critical] habitat of such species.[72]

In the event that an endangered or threatened species is believed to be present in an area, the agency proposing to authorize a project in that area must perform a biological assessment to identify endangered or threatened species likely to be affected by such action.[73]

In 1978, the United States Supreme Court, acting under the Endangered Species Act to protect a species of perch known as the snail darter, enjoined the Tennessee Valley Authority from completing or operating the 100 million dollar Tellico Dam, as the agency was virtually ready to open the gates.[74] In response to that decision, Congress amended the Act to include an exemption procedure. The standard for granting such an exemption is a determination that

i. there are no reasonable and prudent alternatives to the agency action;

ii. the benefits of such action clearly outweigh the benefits of alternative courses of action consistent with conserving the species or its critical habitat, and such action is in the public interest; and

iii. the action is of regional or national significance; and

(B) it establishes such reasonable mitigation and enhancement measures, including, but not limited to, live propagation, transplantation and habitat acquisition and improvement as are necessary and appropriate to minimize the adverse effects of the agency action upon the endangered species, threatened species, or critical habitat concerned.[75]

The standard for an exemption, while stiff, is one that might be met by a tidal power development project if it were part of, or pursuant to, a specific federal energy program or were developed by a governmental entity.

Recent cases suggest that the exemption procedure might never need to be invoked except in the rare case, such as that of the snail darter, where the habitat is the *only* one for the species in question.[76]

Historic preservation concerns, like endangered species and fish and wildlife interests, are a component of the EIS prepared under NEPA, but like these interests they are also protected by more specific acts. The National Historic Preservation Act (NHPA)[77] authorizes the Secretary of the Interior to designate various "sites, buildings, structures and objects significant in American history, architecture, archeology, and culture" for inclusion in a National Register.[78] The inclusion of a site in the Register constitutes general notice of its historical value to a developer whose development would impact the site in question.

The Reservoir Salvage Act[79] which predated NHPA, specifically provides for

> the preservation of historical and archeological data ... which might otherwise be irreparably lost or destroyed as the result of ... flooding ... and other alterations of terrain caused by the construction of a dam by any agency of the United States, or by any private person or corporation holding a license issued by any such agency.[80]

The existence of such sites will not necessarily stop construction of a dam--unless FERC were to determine in its balancing of costs and benefits that the great value of preserving the site proposed to be impacted (probably in conjunction with other benefits) outweighed the benefits to be obtained from energy resource development at that site. In many cases construction of the project would, however, be delayed to allow time for removal and preservation of buildings and or artifacts.

5. *The Coastal Zone Management Act and Other State Controls*

Although the Coastal Zone Management Act (CZMA) is a federal statute,[81] it offers a means for state control, if its application to a tidal power project could survive preemption. The Act, signed into law in 1972, was designed to encourage states to engage in planning and management activities for their coastal zones and to coordinate state and federal programs and activities affecting the coastal zone. The 1976 amendments to the Act[82] instituted a separate system of funding to deal with the impacts of energy development pressures and activities in the coastal zone. This Coastal Energy Impact Program (CEIP) provides grants or loans to coastal communities for, among other things, planning and mitigation measures to deal with environmental and recreational loss associated with energy development.[83]

The continued viability of state coastal management is in some doubt due to cuts in funding, both implemented and proposed, by the Reagan administration.[84] For the present, however, Maine has a federally-approved Coastal Zone Management Plan.[85] Thus, under the CZMA,

> any applicant for a required federal license or permit to conduct an activity affecting land or water uses in the coastal zone of that state shall provide in the application to the licensing or permitting agency a certification that the proposed activity complies with the state's approved program and that such activity will be conducted in a manner consistent with the program.[86]

This means that before issuing permits for the construction or licensing of a coastal energy project, such as a tidal power project, the federal permitting agency would have to receive a consistency certification from the applicant. In addition, the State must concur in this certification.[87] This does not mean that a state has a conclusive veto over federally supported activities in its coastal zone. The Secretary of Commerce may overrule the state and make a determination that in his judgement the activity is consistent or is "otherwise necessary in the interest of national security."[88] In the case of a conflict of opinion between the State and the federal agency, the issue of whether a project is consistent "to the maximum extent practicable" may well be decided in the federal courts.

As for other state controls, as mentioned, the United States Supreme Court decision in *First Iowa Hydro-Electric Cooperative v. Federal Power Commission*[89] held that all state regulatory control over hydro-electric development is preempted by the Federal Power Act. However, a tidal power developer that felt it could meet the state standards might find it more cost effective, and certainly better public relations in the state in which it wished to operate, to apply for state permits rather than challenge the state's authority through the courts. Indeed, the Passamaquoddy Tribal Council, which proposed to construct an 18 MW tidal power project at Half-Moon Cove in Maine's Cobscook Bay, applied for state and local permits simultaneously with its attempts to secure a FERC license.[90]

Under recently passed state legislation, the Maine Waterway Development and Conservation Act,[91] the state permitting process would be streamlined, with only a single state permit required. Any project granted this permit would automatically receive state water quality certification under § 401 of the federal Clean Water Act.[92] Water quality certification is one state control that FERC accepts,[93] providing an incentive for developer submission to the state permitting process. This legislation was part of a larger package known as the Maine Rivers bill that provides for comprehensive planning for the state's rivers, including designating certain rivers and river segments on which no hydro-electric development may take place.[94] FERC, however, has refused to be bound by the plan.[95]

For further discussion of this legislation and other ways in which states might exercise control over hydro-electric development, were the *First Iowa* decision to be modified, see Arnold, "Emerging Possibilities for State Control of Hydro-electric Development."[96] It should be noted that modification of that decision might come through legislation, rather than court decision, as Senator George Mitchell (D.-Me.) has recently introduced legislation that would allow the states to exercise some control over hydro-electric developent through comprehensive planning like the Maine initiative.[97]

Conclusion

Because of the uncertainties created by the still unresolved issues of: 1) FERC preemption under the Federal Power Act, now on appeal to the District of Columbia Circuit in *Monongahela Power Co. v. Alexander*;[98] 2) the extraterritorial applicability of NEPA; and 3) the novelty of tidal power development, the regulatory picture for construction and operation of a United States tidal power generating facility is far from clear. What is certain is that many federal agencies would have input into the process, with FERC being the primary decisionmaker. What is equally certain, based on past experience, is that Maine and the nation have well-organized environmental organizations that would certainly be involved in various stages of the process. In addition, Maine has a history of the formation of project-specific or issue-specific citizen's groups, which would also likely be vocal participants in the permitting process. The recent history for the siting of other energy facilities, such as oil refineries, on the Atlantic coast has not been one of success for the developers.[99] In all events, the regulatory and judicial process for tidal power development in Maine would undoubtedly be long and difficult.[100]

Potential U.S. Regulation of Fundy Tidal Power

A tidal power development wholly within the United States, or a joint Canadian-American venture in Passamaquoddy Bay would be subject to the full range of United States environmental regulation discussed above. It appears, however, that a tidal power project situated entirely in Canada, even if its primary purpose were provision of power to the United States and construction were conditioned on American purchase power contracts, might largely escape United States regulation. Such regulation and environmental review as would occur could be limited to review of the economics of the power purchase agreements and review of the transmission line siting, although there are avenues for possibly obtaining broader environmental review.

1. Federal Regulation

Federal Energy Regulatory Commission (FERC) regulates wholesale ratemaking[101] by utilities and "is responsible for approving agreements between and among utilities for the sale, transmission and exchange of energy and capacity."[102] However, the emphasis in this authority is on the sale of electricity by public utilities within Commission jurisdiction, and the only express authority over a purchase from a utility outside

FERC jurisdiction would be on a request for a wholesale rate increase in conjunction with the transmission of the power purchased.[103] Of course, if the contract involved a concomitant sale of electricity, as such contracts often do, then that aspect of the transaction would also be subject to Commission scrutiny. FERC may hold a hearing in conjunction with such rate filings at which interested members of the public would be allowed to intervene, but the FERC jurisdiction is limited to questions of whether the rates and contracts are "unjust, unreasonable, unduly discriminatory or preferential."[104] These terms are all traditionally defined in the economic sense.

Assuming that a purchase power contract with a Fundy tidal power authority were entered into by a regulated utility prior to construction of the facility and that this contract became part of a rate filing at that time, a novel question would be whether a challenger could claim that approval of the rate would be a "major federal action significantly affecting the quality of the human environment" for which FERC would have to prepare an Environmental Impact Statement (EIS).[105] This approach is problematic for several reasons. It would require that the rate filing occur prior to construction; that the public has knowledge of the filing and its significance; that such persons be allowed to intervene in the rate proceeding; that they then prove that the contract was the *sine qua non* for tidal power construction; and that they succeed in expanding the purview of NEPA[106] to an entirely new area in an arguably inhospitable judicial and political climate for such action. In short the obstacles to this approach are formidable.

The Department of Energy (DOE) regulates the transmission of energy *to* a foreign country under the Federal Power Act with a mandate to assure that exports of power will not "impair the sufficiency of electric supply within the United States or ... impede or tend to impede the coordination in the public interest of facilities subject to the jurisdiction of the Commission."[107] However, there is no comparable authority for the regulation of power imports.[108]

The clearest authority for regulation of the power imported would be Executive Order No. 10485,[109] which provides, among other things, that:

> Section 1.(a) The Secretary of Energy is hereby designated and empowered to perform the following described functions:
>
> (1) To receive all applications for permits for the construction, operation, maintenance, or connection, at the borders of the United States, of facilities for the transmission of electric energy between the United States and a foreign country.

As interpreted, these permit provisions apply to electric power lines constructed for imports, as well as exports. Prior to issuance of the permit the Secretary of Energy is required to make a finding that issuance of the permit is "consistent with the public interest" and to consult with the Departments of State and Defense.[110] If the three Departments cannot agree on the advisability of issuing a particular permit, the matter is to be referred to the President.[111] Conditions may be attached to a permit.[112]

DOE has delegated permit processing to the Office of Utility Systems of the Economic Regulatory Administration (ERA).[113] The regulations promulgated pursuant to the Executive Order[114] caution that in reviewing the application the agency will make a determination of whether an EIS must be prepared, and if so, permit processing will normally take 18-24 months.[115] Applications are reviewed for impacts of the proposed facilities on domestic energy supplies and power reliability, as well as for environmental impacts. The agency commonly prepares EIS's for major transboundary transmission lines[116] in accordance with the Council on Environmental Quality regulations.[117]

The major question that arises is whether any impacts other than impacts of the transmission line construction and power reliability would or could be considered in such an impact statement. The DOE regulations apparently contemplate consideration of power line routing alternatives, rather than consideration of alternatives to construction of the line at all.[118] However, two recent EIS's prepared in conjunction with energy imports indicate that alternative sources of power are considered and their comparative viability and environmental impacts summarily discussed. In both of these cases the only impacts of the proposed import appeared to be from transmission lines, not from the Canadian generating facilities.[119]

Of course, if the tidal power project had already been built at the time of the application for a transmission line permit, questions of tidal power impacts would be moot. Again, only if the permit were required early enough in the process to catch the environmental implications of United States investment in the Fundy project and influence the project's viability, would a NEPA review be likely to address coastal ecosystem effects of the Nova Scotian project itself. This is a possibility if guaranteed access to the New England electricity markets is a prerequisite to construction of the project, and if construction of a new power line is a prerequisite to assured access.

This may indeed be the case. The Tidal Power Corporation's 1982 update of the economic status and financial feasibility of the Fundy proposal was based on an allocation of 10% of the energy produced to the Maritimes and 90% to the United States (45% each to New York and New England).[120] That report states in its conclusion No. 4 that "iron clad contracts for the sale of output would be a necessary element in the security required by lenders...."[121] The report also assumes the necessity of construction of new high voltage transmission lines from Cobequid Bay to delivery points in Ludlow, Massachusetts and Pleasant Valley, New York.[122] Thus U.S. investor utilities might well have to seek a presidential permit prior to construction of the Fundy project in order to provide the assurances that the Tidal Power Corporation requires. Even if construction of an entirely new transmission link were not required and only an upgrading of the existing 345 kV Maine-New Brunswick line were planned,[123] such upgrading would also require DOE approval, although perhaps not a full application process.[124] If the timing for seeking such a permit or approval were preconstruction, this would be the best handle for getting NEPA review of the environmental effects of the tidal power project itself, and not just the transmission lines.[125]

Under these circumstances there would appear to be a colorable claim for NEPA consideration of the effects of the construction and operation of the tidal power project, as well as of the transmission links, although the claim may suffer from its novelty. Novel arguments may also surface in opposition to any NEPA suits (e.g., to expand the scope of the EIS), such as the argument presented by the Power Authority of the State of New York in *Greene County Planning Board v. Federal Power Commission*[126] that the Planning Board's petition for review of the granting of a presidential permit for failure to comply with NEPA "presents a non-justiciable political question, *viz.*, whether the permit was issued in accordance with the proper conduct of the foreign relations of the United States."[127] (In fact, the NEPA question was preempted in that case by Congressional legislation specifically exempting the transmission facilities in question from NEPA requirements.)[128]

Assuming NEPA review did encompass consideration of the effects of the tidal barrage, the Secretary of Energy could, of course, deny the permit if he found the granting of the permit and the concomitant effects of the construction and operation of the tidal power dam would not be "consistent with the public interest." He could also grant the permit, but attach conditions that in turn would impact the feasibility, design, or operation of the tidal power project itself, pursuant to a contract relying on that conditional permit.[129]

The state in which a boundary crossing occurs--in this case Maine--has an opportunity to participate in the EIS review process through consultations. If the effects of the tidal barrage and generating facility were determined to be within the scope of the EIS, then arguably the state would also have regulatory control under the federal Coastal Zone Management Act (CZMA) consistency requirements.[130] Consistency with Maine's program is determined by obtaining all applicable permits under the "core laws" that constitute the state's coastal zone management program, the most prominent of which, here, would be the Site Location of Development Act.[131] An interesting question is whether a federal determination that approval of the transmission line required examination of impacts of the Canadian power project would push the state into expanding its site location review to consider such impacts as well. As the discussion below will indicate,[132] the state statute probably would not be interpreted to encompass consideration of the transboundary impacts of the tidal facility when transmission line siting was under review; but the federal action, if it preceded the granting of the state permit, could affect this interpretation.

2. State Regulation

The primary state regulatory agencies in Maine[133] with jurisdiction over transmission lines are the Public Utilities Commission and the Board of Environmental Protection.

The Maine Public Utilities Commission must give prior approval both for the erection by any electric company of any transmission line carrying 100 kilowatts or more[134] and, since 1982, for the purchase of "any right, title or interest in generating capacity, transmission capacity or energy." "'Energy' is defined to include an entitlement to energy for a period greater than 3 years."[135] In both cases, the Commission's order granting approval must contain specific findings with regard to the need for the purchase contract or transmission line construction in question. If a need is found to exist, the Commission "shall issue a certificate of public convenience and necessity"[136] for the purchase or construction of the proposed facilities, as the case may be.

Review of the purchase power contract itself would give the Commission an opportunity to inquire into such issues as the reliability of the power to be purchased and the availability of lower cost alternatives. Also, if the purchase of power could expose the Maine utility to substantial legal liability, this could be a consideration in review of the contract. This possibility may be remote, but one Tidal Power Corporation (TPC) proposal, indeed the one considered financially preferable, is for a joint venture with an operating company, equity holders, and participating utilities. The utilities would provide "construction capital in return for rights to purchase plant output at operating cost plus an agreed on return on equity."[137] Such a structure would arguably expose the participating utilities to liability for any damages caused by the operation of the tidal power plant, a factor that would go into the PUC calculus in determining whether the public convenience and necessity would be served by approving the purchase power contract.

The PUC lacks authority, however, to consider environmental effects in the process of approving applications for transmission line construction. According to judicial interpretation of the relevant statutes and legislative history in a 1974 case, *In re Bangor Hydro-Electric Co.*,[138] this authority is allocated to the Board of Environmental Protection (BEP) pursuant to Maine's Site Location of Development law (discussed below).[139] Thus the PUC's decision would focus on questions of safety, reliability, and economic reasonableness of the proposed facility. The legislation governing PUC authority does provide that in the event the BEP modifies proposed facilities for which the PUC has already granted a certificate, the company proposing the project will notify the PUC of both the changes and necessary cost adjustments. If those adjustments "exceed the cost relied upon by the commission ... by more than 20% of the original cost," the Commission shall reopen the proceedings and reconsider the question of need.[140]

It should be noted that a public hearing is required prior to the issuance of a certificate of public convenience and necessity.[141] Politically sophisticated citizens and public interest groups opposed to new transmission lines and/or to the underlying tidal power project primarily on environmental grounds might also investigate and argue the lack of a "need" for the proposed facilities because of economic or reliability concerns or lack of need for additional power supplies. In addition, local opposition to the power line itself could be significant,[142] particularly if its primary purpose is to serve the markets in southern New England and New York.[143] Interestingly, the Tidal Power Corporation identifies "the ability to deny transmission facilities [and] [t]he ability to withhold regulatory approval of purchase contracts" as *de facto* United States vetoes on tidal power development.[144] The question is whether under the standards applicable to these approvals the state regulatory authorities would have a basis for denying the permits. Of course reviews of power purchases in states other than Maine would be involved as well.

Local transmission line siting permits are also required from the Department of Transportation for siting lines along most public highways; from municipal officers for siting along city streets or town ways and certain state and state aid highways; and from county commissioners for highways not within the jurisdiction of the other agencies.[145]

Numerous other state or local permits may be required, such as compliance with shoreland zoning requirements[146] or, if the transmission line passes through the unorganized territories, with legislation and regulations of the Land Use Regulation Commission;[147] but the primary regulatory agency is the Board of Environmental Protection. [148]

The Maine Site Location of Development Act, originally enacted in 1969, is Maine's most comprehensive environmental review statute. The purpose of the Act is to provide a mechanism for the State "to control the location of those developments substantially affecting local environment in order to insure that such developments will be located in a manner which will have a minimal adverse impact on the natural environment of their surroundings and protect the health, safety and general welfare of the people."[149] The Act defines a "development which may substantially affect the environment" to include, among other things, any development "which occupies a land or water area in excess of 20 acres."[150]

This definition would certainly cover a major new transboundary transmission line. Indeed, the statute explicitly assumes coverage of a power generating facility of more than 1000 kW or a transmission line carrying 100 kV or more "proposed to be erected within this State...." Such developments must obtain a permit from the State BEP prior to commencing construction. The Board may, in its discretion, hold a hearing on a permit application, and probably would for a major new transmission line, especially if members of the public requested one. The standards for approval of a development are that the developer demonstrate that 1) he has the financial capacity and technical ability to meet State pollution control standards (air, water, solid waste disposal); 2) he has made adequate provision for traffic movement; 3) he has made adequate provision for fitting the development harmoniously into the existing natural environment without adverse effect on existing uses, scenic character or natural resources in the area; 4) the soil types are suitable; and 5) ground water resources will not be at risk. In addition, the statute has a specific requirement that

> in the case of such transmission line or pipelines, [the board] shall consider whether any proposed alternatives to the proposed location and character of such transmission line or pipeline may lessen its impact on the environment or the risks it would engender to the public health or safety, without unreasonably increasing its cost.[151]

The regulations promulgated under the Site Location law flesh out the third standard, specifying a finding of "no adverse effect on the natural environment" with fifteen separate considerations ranging from effect on air quality to preservation of historic sites to effects on climate alteration.[152] While the language of the Act seems to focus on impacts of the individual project--here the transmission line and corridor itself--the regulations specifying the general scope of review under the statute are potentially broad enough to allow inquiry into the impacts of the tidal power dam itself. That section provides:

> In reviewing applications for approval of proposed developments under the Site Location Law, the Board shall consider the size, location, and *nature of* the proposed development in relation to:
>
> A. The *potential primary, secondary and cumulative impacts* of the development on the character, quality, and uses of land, air, and water on the development site *and on the area likely to be affected by the proposed development*; and
>
> B. *The potential effects* on the preservation of *the public's* health, safety, and *general welfare*.[153]

It is unlikely, however, that a regulation with no clear grounding in the statute will be broadly interpreted by the state to form the basis for examination of the impacts of Fundy tidal power. A review of the relatively small number of reported cases interpreting the statute reveals none that address the question of secondary impacts, i.e., impacts from necessary, related developments. Furthermore, unlike federal agencies in NEPA reviews, the State Board apparently does not have authority to allow consideration of alternative methods and sites to proposed projects.[154] Although the Board would, under specific provisions of the statute in question, be required to consider alternative sites for the transmission line,[155] it probably would not look into the question of whether there were less environmentally destructive alternatives to provision of power than importation of tidal power.

That the effects of a tidal power dam in Canada is not an issue contemplated for consideration in power line siting does not mean that it would not be raised. Again the timing of the application for the permit would be critical. If the dam had already been constructed prior to application, only impacts of the line itself would be considered. But if the permit were needed for a preconstruction purchase power contract, then the public hearing process might attract opponents of the tidal dam construction.

The location of public hearings could be a factor in development of this issue. If the power line were to be sited far from the coast with the hearings held in localities close to the transmission corridor,[156] this could affect awareness and participation by parties potentially concerned about effects of the tidal dam, e.g. coastal property owners and fishermen. However, at least one of the State's major environmental groups is obviously monitoring developments and would be likely to participate in whatever public fora are made available.[157]

A 1979 analysis of "New England/Maritime Region Electrical Interconnections" prepared jointly by New England and New Brunswick utilities concluded that

> The large number of regulatory approvals required give indication that a long lead time may be necessary to construct additional 345 kV transmission lines within the State of Maine. The time required to gain permission from the large number of state agencies and local governments which must grant approval will make the addition of transmission line capacity between the United States and Canada a lengthy process.[158]

Conclusion

Although considerable state legislation and a federal permitting process exist to examine impacts from a major new transmission line and the siting thereof, the extension of these mechanisms to examine the environmental effects within the United States of the tidal power project itself, when located wholly within Canada, is a largely untested proposition. The existing National Environmental Policy Act should be interpreted to include examination of these impacts, especially, where, as here, United States participation is the *sine qua non* of the project. However, this requires an opportunity to look at the project prior to construction of the tidal power barrage. The financial requirements of the Fundy project, with the consequent necessity (probably) for United States transmission permits in hand prior to construction, make it likely that the early opportunity for environmental review will occur in this case. But this is circumstantial. If the financial situation were different, or if existing transboundary transmission facilities were adequate to accomodate tidal power energy outputs, there might be no regulatory handle to investigate the transboundary effects of this project.

This possibility suggests that the United States environmental legislation is inadequate. While there has been substantial investigation of our obligations under NEPA and, more recently, under Executive Order 12,114[159] to consider the impacts of United States activities abroad, there has been less concern with what recourse we may have when environmental effects are imported. Were there no United States participation in the project, then the issue would not be one for domestic legislation, but for resolution in the international arena through traditional diplomacy, treaties, and international courts, or by reference to a transnational body, e.g., the International Joint Commission.[160] Here, however, if the federal and state regulatory agencies permitted the transmission lines and the Maine PUC permitted the purchase of energy, then, under international law, the United States and the state might be deemed to have consented to any damage that occurred from the construction and operation of the tidal power facility.[161]

In the present case the current leadership of the Tidal Power Corporation has indicated that the Canadian environmental review process will examine United States impacts[162] and that it is essential that a compensation system be in place prior to project construction.[163] But United States participation in the project should trigger a full domestic environmental review prior to conclusion of the contract for the power purchase. If Canada, or Mexico, is constructing large energy facilities primarily or solely for export purposes, and near enough to our boundaries to have substantial impacts, then the United States laws should assure the examination of those impacts before allowing energy imports.[164]

Such a policy could be adopted in the narrowest case by federal legislation giving the Department of Energy the authority to review contracts for the importation of electricity,[165] just as they currently review exports, and by subjecting such importations to the requirements of NEPA where the contract was sought by the foreign utility prior to construction (or even simply prior to the operation) of the proposed energy project. A proposal for such legislation was submitted to the Senate Committee on Environment and Public Works following their hearing on Fundy tidal power. (It is reprinted in an addendum to this chapter's notes.)

Such legislation would not, of course, preclude cooperative environmental review, which obviously should be pursued in cases such as Fundy tidal power where a transboundary region will be affected and where financial investments will be made and benefits accrue on both sides of the border. Efforts to establish such cooperative review are already under way, including those under the auspices of the New England Governors' Conference/Conference of Eastern Canadian Premiers.

Such cooperative efforts are essential. They complement the proposed domestic reforms, but do not preclude the need for improvements in the scope of domestic environmental review to assure that, to the extent possible, it encompasses consideration of the transboundary effects of energy developments abroad, where power will flow into United States grids.

The Canadian Decisionmaking Framework

The Canadian legal framework for managing tidal power development may be described as a loosely knit and potentially tangled net. Jurisdictional tangles may occur at three levels -- federal vs. provincial, because of constitutional uncertainty, federal vs. federal, and provincial vs. provincial, because of overlapping departmental mandates. Government decisionmakers, lacking clear legislative policies or guidelines and not subject to intensive judicial scrutiny, possess great discretion in requiring and weighing scientific data. This section summarizes the Canadian legal system from five perspectives: constitutional powers, federal roles, provincial roles, legislative open-endedness, and lack of judical review.

Constitutional Powers

The Constitution Act, 1867, the document which divides powers between the federal and provincial governments, draws no clear jurisdictional line for controlling tidal power development. The Act mentions water power only once and merely grants the federal government jurisdiction over water power from canals. The Act does not explicitly mention hydro-electric development of rivers or estuaries, does not clearly address jurisdictional rights over marine waters or the seabed, and remains silent about who should control environmental effects of technological developments.

As a result, it remains for the provinces and federal government to define their own jurisdictional lines based on a number of tangential constitutional provisions. The federal government may claim the right to regulate tidal power development through eight major sources of legislative power granted by the Constitution Act, 1867: power over Spending, the Regulation of Trade and Commerce, Navigation and Shipping, Sea Coast and Inland Fisheries, Extraprovincial Works and Undertakings, Works Declared to Be For the General Advantage of Canada or the Provinces, the Treaty Power, and the federal residual power for the "Peace, Order and Good Government of Canada."[166]

The provincial governments may rely on at least five constitutional powers: power over Property and Civil Rights in the Province, Local Works and Undertakings, All Matters of a Merely Local or Private Nature, the Management and Sale of Public Lands, and Electrical Energy Production. Section 92A, a recent amendment to the Constitution Act, has apparently endowed the provinces with broad powers over electrical energy

> [I]n each province the legislature may exclusively make laws in relation to ... development, conservation and management of sites and facilities in the province for the generation and production of electrical energy.

Such a provision could be a source of constitutional controversy concerning Fundy tidal power since a number of unanswered questions are embedded within this language. Will the Fundy tidal power site be "in the province" for purposes of constitutional jurisdiction? Since ownership and jurisdiction over areas of the east coast offshore seabed is still open to question, the federal and provincial governments might disagree on the answer. If the facility is considered to be within the province, what is the meaning of the words "conservation and management of sites and facilities'? Do the words "conservation and management" encompass powers over environmental conditions or are the words more limited in scope, referring to economic, physical and technical aspects of development such as deciding the rate of electrical energy production and the location and types of electrical facilities?[167] If "conservation and management" includes power over regulating environmental effects, how far does such power extend? Just at the site/facility proper? Or at the site/facility in terms of a larger area --extending, for example, to include the head pond and perhaps a seaward zone, as well?

This at least theoretical potential for controversy will not necessarily lead to conflict in practice. Indeed, the practice of the federal and provincial governments shows a trend towards cooperation over resource management. In 1982 the federal government and the Province of Nova Scotia set aside the question of seabed ownership by entering into an oil and gas resource sharing agreement.[168] For the Annapolis Tidal

Power Project described in chapter 1, the Nova Scotia Department of the Environment formed an Intergovernmental Environmental Advisory Committee with members from the Nova Scotia Department of Environment, the Nova Scotia Department of Agriculture, the federal Department of the Environment, the Department of Fisheries and Oceans, and the Atlantic Geoscience Centre. The Committee continues to monitor and advise on environmental matters even after construction of the tidal barrage. For the Fundy tidal development proposed in 1977, a federal-provincial environmental assessment panel was formed which drew up draft guidelines for an environmental impact statement. Such guidelines have never been acted upon, however, due to altered development plans. Both Nova Scotia and New Bruwnsick have special federal-provincial committees which advise concerning stream alterations.

Federal Roles

In the event of a direct conflict between federal and provincial legislation relating to environmental aspects of tidal power development, federal legislation would likely prevail, according to the judicially created doctrine of paramountcy, so long as the federal government was acting under the cloak of its own "exclusive" legislative power such as the power over seacoast and inland fisheries.[169]

Five federal entities -- the Department of the Environment, the National Energy Board (NEB), the Department of Fisheries and Oceans, the Department of Transport and the Department of Energy, Mines and Resources -- will likely play substantial roles in controlling tidal power development.

The Department of the Environment would likely oversee the Environmental Assessment Review Process (EARP). The process, established by Cabinet directives in 1973 and 1977 and strengthened by a Cabinet Order in June 1984,[170] is mandatory for all projects initiated by federal departments or agencies (except certain Crown corporations),[171] for projects requiring federal funds or lands, and for proposals having an environmental effect on an area of federal responsibility. The process has evolved into essentially a three-stage review. First, the initiating federal department (that is, the department with the lead involvement in a development project) undertakes an initial environmental screening to establish the need for an environmental assessment. Second, if the department decides a proposed project carries the potential for significant environmental effects, the department undertakes (with the assistance of the Department of Environment, if required) an initial enviromental evaluation based upon available information. Third, if after an initial evalution the department decides a full environmental impact assessment (EIA) should be undertaken, the development proposal is referred to the Minister of Environment for public review by an Environmental Assessment Review Panel, usually consisting of public servants and independent experts. A referral may also occur if public concern makes a public review desirable. The Panel would review the EIA, hold public hearings, and make recommendations to the Minister of Environment and the Minister of the initiating department who would then confer with other relevant departments or agencies over the extent to which the recommendations should become mandatory. For the first two stages of review, every initiating department is required to develop a list of the types of proposals not producing any adverse environmental effects and automatically excluded from formal panel review, and a list of the type of proposals producing significant adverse environmental effects and automatically referable to the panel process.

The National Energy Board would have to grant a certificate of public convenience and necessity before construction and operation of an international power line could commence, and would have to issue an export license authorizing tidal energy export to the United States market. As preconditions to such permits, the Board may require environmental reviews.

The Department of Fisheries and Oceans, pursuant to powers granted by the Fisheries Act,[172] could require fish-passes and adequate water flows at the tidal barrage and could require extensive scientific studies concerning fish habitat alterations. The Minister of Fisheries and Oceans could also modify construction plans and even prohibit (with Cabinet approval) the project altogether.

The Minister of Transport, through the Navigable Waters Protection Act,[173] has three major controls over tidal power development. The Minister would have to approve and could condition tidal barrage plans and locations. The Minister would have to approve the dumping of fill and the dredging of materials from the seabed, and he or she might require certain water levels and flow rates at the tidal site in order to protect navigation.

The Department of Energy, Mines and Resources could play a central role in tidal power decisionmaking at two levels. First, the Department might financially support the project and thereby indirectly influence how and when tidal development should proceed. (The Department, together with the Department of Industry, Trade and Commerce, contributed $25 million to the Annapolis development out of a projected cost of approximately $46 million.) Second, the Department might influence tidal development through its mandate to formulate national energy policies. For example, if the Department decided a second nuclear reactor at Point LePreau, New Brunswick, was in the national interest and financially supported the project, Fundy tidal power might become economically infeasible.

Provincial Roles

Nova Scotia tidal power development would primarily involve the interaction of three provincial entities. The Tidal Power Corporation, funded initially by $25 million in provincial money and headed by a Board of Directors and Chief Executive Officer appointed by the Nova Scotia Cabinet, would be the chief promoter. To facilitate tidal development, the Corporation is granted four major powers. It may hire an extensive staff of consultants and technical advisors to undertake engineering and environmental studies. It may agree with U.S. federal or state governments, U.S.

companies, U.S. residents or their Canadian counterparts to undertake independent or joint studies relating to tidal power. It may expropriate land and hold title to Nova Scotia lands, water, and air. And it may enter into contracts relating to tidal development such as sales contracts, distribution contracts and construction contracts.[174]

Environmental regulation would be shared by two provincial departments. The Department of Environment, through power conferred by the Water Act,[175] would have to approve any hydro-electric power plans, could require environmental information as a precondition to approval, and could attach conditions to any water authorization, for example, requiring minimal or maximum water levels, fish passage, and adequate compensation to persons injured by tidal alterations. The Department of Lands and Forests, with Cabinet approval, may grant and place conditions on the use of right-of-ways for power transmission lines. The Cabinet, upon recommendation of the Minister of Lands and Forests, has the authority to include environmental or socio-economic conditions in a lease of Crown lands and the associated right of water flowage for storing and developing water power.[176]

New Brunswick legislation concerning water power development is very similar to Nova Scotia's framework in providing for both promotion and protection. The New Brunswick Electric Power Commission, a Crown corporation, may finance electric-energy projects, may purchase or expropriate land, water and water power, may construct water power facilities including dams and transmission lines, and may flood lands with or without landowner consent.[177] Pursuant to the Clean Environment Act,[178] the Minister of the Environment (recently renamed the Minister of Municipal Affairs and Environment) may require hydro-electric proponents, before undertaking a project, to submit plans and information and to attain a permit.[179] The provincial Cabinet is authorized to regulate any facet of hydro-electric development, including location, construction, water flow, and water storage.[180] Pursuant to the Crown Lands and Forests Act,[181] the Minister of Natural Resources (recently renamed the Minister of Forests, Mines, and Energy) may impose conditions for habitat protection in any lease of water and land for a tidal barrage and water storage, or in any right-of-way for power transmission lines.

Legislative Open-Endedness

Canadian legislation contains numerous discretionary provisions leaving decision-makers largely unassisted by Parliamentary-ordained procedures, schedules, or policy guidelines.

The National Energy Board Act[182] allows the National Energy Board (NEB), in issuing a certificate of public convenience and necessity or export license, not only to consider all matters it considers relevant but also to place environmental, social, and technical conditions into such permits. Broad language governing certificate issuance states: "The Board may issue a certificate subject to such terms and conditions as the Board considers necessary or desirable in the public interest."[183] Similarly, language governing license issuance states: "Subject to the Regulations, the Board may, on such terms and conditions as it may impose, issue licenses...."[184] No set schedule binds the Board in holding hearings or reaching decisions. Whatever procedures are mandated may be omitted by Board discretion as set forth in the Rules of Practice and Procedure: "The Board may, in any proceeding before the Board upon an application, direct either orally or in writing that ... these Rules ... shall not apply ... for the purpose of ensuring the expeditious conduct of the business of the Board."[185]

NEB policies for requiring and weighing environmental information are minimal. In a 1972 report to Cabinet, the NEB stated its environmental responsibilities for export licensing were two-fold: to assure power production for export would not exceed pollution limits set by other government agencies and to weigh export benefits against adverse environmental impacts in order to ascertain whether there was a net advantage to Canada.[186] Board regulations governing export licensing set out only a skeletal requirement for environmental review. The applicant is required to produce "evidence on any environmental impact that would result from the generation of power for export."[187] Board Rules of Practice and Procedure, governing tranmission line certification, phrase informational requirements in general terms. An applicant must provide "an assessment of the probable environmental impact of the international power line, including a description of the existing environment and a statement of the measures proposed to mitigate the impact."[188] NEB Environmental Guidelines for International Power Lines are flexible, allowing the applicant to provide information commensurate with the scope of the project.[189]

NEB review of marine environmental effects of facilities constructed for energy export remains uncertain in timing and scope. Although project proponents generally do seek NEB approval early in the planning process, legislation does not require export license approval before construction begins. The Board has waivered concerning the extent of jurisdiction over marine environmental effects of industrial developments. The NEB, when approving the LNG terminal at Lorneville, New Brunswick and the gas pipeline to the United States, expressed concern over safety of marine operations on the Bay of Fundy but concluded that an assessment of such factors as the docking of LNG tankers and adverse weather conditions was beyond Board jurisdiction and better left to Termpol Review (an interdepartmental review headed by the Canadian Coast Guard).[190]

In a more recent decision, however, the Board in effect assumed broad jurisdiction over marine matters. In the Trans Mountain Pipe Line Company case,[191] Trans Mountain Pipe Line Company sought a provisional certificate of public convenience and necessity for a pipeline to carry oil from a proposed terminal at Low Point, Washington through Canada to the landlocked northern states of Montana, North Dakota, Minnesota, and Wisconsin. Since Trans Mountain had not carried out environmental studies on the effects of extra tanker traffic and a U.S. marine terminal on Canadian marine waters and coastline, and since the NEB was unsure U.S. regulatory agencies would adequately consider the effects of a U.S. terminal on Canada, the

Board adjourned the hearings and suggested Trans Mountain apply for a rehearing after undertaking further environmental studies. The NEB subsequently approved the pipeline proposal subject to numerous conditions such as meeting all U.S. regulatory requirements and implementing environmental mitigation measures.[192]

Environmental review by the Department of Fisheries and Oceans, pursuant to the Fisheries Act, is also open to official discretion. Although the Act authorizes Cabinet to pass regulations governing departmental reviews, no regulations have been passed to date and thus the Minister is free to set his or her own review schedule, informational requirements, and public consultation policy. The applicant has no pre-established right to confront (for example, through cross-examination) those opposed to the project application but would have to rely on the general right of "a reasonable opportunity to make representation." Departmental formulation of a fish habitat management policy could incorporate more specific review procedures but has not moved beyond the draft stage.

The Environmental Assessment Review Process (EARP) operates with only slim legislative guidance. The Government Organization Act, 1979,[193] merely requires that the Minister of Environment ensure new federal projects are assessed early in the planning process for potential adverse environmental effects, that a further review be carried out for projects likely to have significant adverse effects and that the results be taken into account. No formal review mechanism has existed to assure EARP recommendations are implemented by the initiating federal department.

A recent Cabinet Order, however, will likely expand the procedural safeguards for public review and post-review implementation of recommendations. Panels are required to develop detailed operating procedures and to allow the public access to and sufficient time to examine submitted information prior to a public hearing.[194] The Federal Environmental Assessment Review Office (FEARO) is required to provide written procedures to ensure procedural and policy consistency among various public reviews.[195] FEARO has subsequently issued basic operational procedures (such as requiring EARP panels to make reports by technical specialists available to the public) and rules for public meetings (such as requiring reasonable notice of public meetings).[196] The initiating federal department is required to establish suitable implementation and monitoring programs for EARP recommendations applied to a project proposal.[197]

The Minister of Transport, in approving construction proposals pursuant to the Navigable Waters Protection Act, is not bound by established procedures or a fixed schedule -- not even a general exhortation such as "the proponent shall have a reasonable opportunity to be heard." The Minister is free to impose conditions in permits "upon such terms ... *as he deems fit*."[198]

Provincial legislation and regulation also grant considerable discretionary powers to government officials. In Nova Scotia, the Water Act gives the Minister of Environment broad power to impose conditions in water use authorizations. The Lands and Forests Act grants the Minister of Lands and Forests and the Cabinet free rein to impose conditions on land leases and power transmission line right-of-ways. In New Brunswick, regulations pursuant to the Clean Environment Act allow the Minister of the Municipal Affairs and Environment to waive permit requirements altogether for alterations to tidal waters[199] or in the case of an emergency as defined by the Minister, to issue a provincial permit allowing the project to proceed before studies and plans are complete.[200] Water alteration permits may contain whatever terms and conditions the Minister desires.[201] The Crown Lands and Forests Act grants the Minister of Forests, Mines and Energy wide discretion to impose conditions in any lease of land and water and in any power transmission line right-of-way.

Restrictive Judicial Review

Given the legislative delegation of broad discretion to administrative decision-makers, Canadian courts have tended to be restrictive in providing judicial review. For example, when various environmental groups challenged the National Energy Board's grant of a certificate to construct an oil pipeline from Norman Wells, Northwest Territories to Zama, Alberta, the court responded to the charge of deficiencies in social and environmental considerations with a hands-off attitude:

> The Court is not entitled to substitute its opinion as to whether or not these facts justified a finding of public convenience and necessity for the opinion of the Board. In a situation of this kind, the determination of public convenience and necessity is not a question of fact, but is, rather, a formulation of an opinion by the Board, and by the Board only.[202]

In a similar vein, when the Township of Tosorontio requested a court to require an EARP hearing for a proposed radioactive storage site, the court avoided confrontation with the administrative agency by finding that

> While the apparent failure to follow the procedures so glowingly advertised by the Minister of the Environment in his May, 1979, publication, *Revised Guide to the Federal Assessment and Review Process*, may be regarded as peculiar in such a conspicuous situation, the Guide created no legal obligations.[203]

Canadian judges could take a more active role in supervising the fairness of administrative procedures and perhaps even undertaking a substantive review of the context of environmental assessments. In formal federal administrative processes such as the quasi-judicial hearings of the NEB, the Federal Court Act[204] not only allows the Federal Court of Appeal to review administrative decisions or orders for failure to observe the open-textured principle of natural justice or for an error of law, but also to judge whether a decision is made arbitrarily and capriciously.[205]

For the more informal federal environmental assessment review process, the Environmental Assessment and Review Process Guidelines order[206] could be interpreted by a court to provide a legal obligation for holding formal public reviews where a federal-related project carries significant adverse environmental implications. The Order requires public review when a project falls under a list of the types of proposals producing significant adverse environmental effects, when the environmental effects of a project are unknown and not subject to further study or reassessment, and when public concern makes public review desirable.[207]

However, a person seeking judicial review of EARP proceedings could face major difficulties. The extent to which the assessment procedures are legal obligations appears to remain in somewhat of a legal twilight zone, for although the Cabinet Order provides an additional legal underpinning, the Order itself still refers to the provisions as "Guidelines". A person seeking to challenge the EARP process would be required to have legal standing, a judicially created requirement with uncertain parameters, but which could demand a special personal interest in the proceedings, such as a property interest, which might be negatively affected by a proposed project.[208] No special statutory right of review exists for the EARP process, therefore, review would take place according to the Federal Court Act.[209] When an administrative tribunal undertakes only investigative and recommending functions and does not make final decisions, only minimal procedures are likely to be required, particularly when the process is intended to be informal.[210] Regarding the decisions of Ministers and Cabinet on whether to accept EARP recommendations, courts have been reluctant to interfere with the policymaking, legislative functions of the executive.[211]

The Canadian Charter of Rights and Freedoms also offers a legal route for seeking judicial review of the decisionmaking process for a project proposal. The Charter establishes individual rights which government may not violate, and if the rights are violated, a court may grant a broad range of remedies including a declaration that an administrative decision (for example, a decision to grant or deny a license or permit) is of no effect or an order requiring an administer to take appropriate action (for example, to provide an oral hearing). To apply the Charter in the environmental field, however, would require overcoming a number of legal hurdles. One would have to establish the existence of a legal right protected by the Charter. Section 7 of the Charter provides everyone the "right to life, liberty, and security of the person and the right not to be deprived thereof except in accordance with principles of fundamental justice." Creative judicial interpretation would be required to extend the meaning of "life, liberty and security of the person" to include environmental rights such as the right to a clean environment, the rights to health and bodily integrity, or the right to a livelihood from renewable resources. Even if an environmental right were recognized, Section 1 of the Charter would still allow the government to argue that any restriction on or interference with the individual's right was "justified in a free and democratic society."[212]

Canadian and U.S. Decisionmaking Frameworks: A Comparison

The Canadian and United States decisionmaking frameworks for controlling tidal power development display at least three major similarities. Both systems involve overlapping jurisdiction between the federal government and state or provincial regulators. Both frameworks carry potential for interagency conflicts, for example, in the United States among the Department of Energy, the Environmental Protection Agency, and the Army Corps of Engineers, and in Canada among the National Energy Board, Environment Canada, Transport Canada, and the Department of Fisheries and Oceans. Both systems also possess energy agencies which display a preference for licensing transboundary energy exchanges. The U.S. Department of Energy, though it would be responsible for preparing the EIS and would receive input from environmental agencies such as EPA and the U.S. Fish and Wildlife Service, is primarily an energy promotion agency with a policy of favoring Canadian imports. The National Energy Board likewise favors exports to the U.S. Both agencies are likely to downplay environmental concerns. For example, the Board rejected Department of Environment proposals to condition its approval of power exports from a coal-fired plant near the U.S.-Canadian border on the installation of scrubbers and on compliance with existing provincial air quality standards. Neither the DOE nor the NEB is likely to go beyond existing environmental standards, if doing so would limit energy exchanges that both agencies favor, even if transboundary environmental issues are present. Both will rely heavily on specific state, provincial, and federal permits for environmental conditions.

Two major differences occur in the preparation of Canadian and United States environmental impact statements. The Tidal Power Corporation, if it is the ultimate proponent of the project, would prepare the impact statement. In the United States, the permitting agency, for example, the Department of Energy or its contractor, prepares the EIS, although this is usually based on an environmental assessment prepared by the project proponent.

While the Canadian EARP panel, specially constituted for the project under consideration, would formulate regulations for EIS preparation specific to Fundy tidal power, DOE would not establish such an *ad hoc* panel for review. Rather, it would rely on existing personnel and on existing regulations of DOE and the Council on Environmental Quality (CEQ) for impact statement preparation. However, in recognition that each project has unique features, DOE would conduct a "scoping" process for preparation of the EIS. Under CEQ regulations, the scoping process involves the public and all interested federal and state agencies in the early stages of identifying potential impacts and alternatives. This process is functionally analogous to the drawing up of specific EARP guidelines for impact statement preparation for an individual project.

Perhaps the greatest divergence between the decisionmaking systems exists, however, in the area of judicial review. Controversies, either between the states or provinces and federal agencies or between proponents and opponents of projects, are much more likely to reach the courts in the United States. In Canada such questions are much more likely to be settled outside of the court system, through negotiations among the parties. The Canadian agencies arguably have more freedom to negotiate, in that they can make concessions or impose conditions within their broad discretion that might be prohibited by the statutes and regulations in the more codified system in the United States. When controversies do reach the courts, both Canadian and U.S. standards of judicial review prohibit the court from substituting its judgment for that of the administrative decisionmaker. Because administrative judgments are more circumscribed by regulation in the United States, however, the courts there are more searching in their review of administrative decisions and apparently are less deferential than Canadian courts to decisions of federal departments.

Concerning the question of which decisionmaking system -- the highly structured U.S. system or the highly flexible Canadian system -- is preferable for addressing uncertain environmental effects, the answer may, at first glance, appear to favor the U.S. approach. Extensive judicial review promises to provide an objective "second look" at the scientific record to identify knowledge gaps and to assure physical, biological, chemical, and social data bases are developed as far as needed and practically feasible. A clear and detailed procedure for the decisionmaking process promises to grant all parties fair participation and full information.

Three basic realities, however, caution against such an initial impression. First, even in a highly flexible system, decisionmakers may, in fact, require a fully developed scientific information base and implement fair participatory procedures. Second, the U.S. adversarial system may not only needlessly delay beneficial projects which do not bear major environmental consequences, but by fostering a win-lose situation may hinder constructive compromise and a balancing of public interests. Third, no legal or administrative structure can guarantee "good" environmental decisions, for in the end many questions are not "hard science" questions involving physical, biological or chemical effects or technological feasibility, but are "personal value" questions involving the ranking of preferences -- social, economic, aesthetic and environmental. Reasonable persons, depending on their interest in a particular project, will vary greatly in their preferences. An industrialist or unemployed worker will likely rank economic considerations higher than environmental, while a naturalist would likely reverse the order. Reasonable persons will also likely differ in opinions as to "the best" decisionmaking methodology. Some may favor an adversarial process with extensive procedures and broad judical review in order to assure full consideration of environmental effects and perhaps even to delay an undesired industrial project. Others may favor a discretionary process where all parties cooperate within a flexible timeframe to negotiate mitigative measures. Government decisionmakers, listening to the conflicting public views, face the difficult task of striking a balance among values.

As this chapter has shown, both the Canadian and U.S. decisionmaking frameworks contain numerous catch-points where government officials may alter the course of a Fundy tidal power project. Whether decisionmakers eventually choose to use such mechanisms to severely restrict or prohibit tidal power development will likely depend on two key factors -- the sufficiency of scientific information and the strength of public expressions for and against harnessing the tides of Fundy.

Notes

1. The section of this chapter describing the U.S. legal framework is an excerpt from Massey, "Tidal Power Development: Environmental Decisionmaking in the United States," 3 Can.-Am. L.J. (1984), reprinted with permission of the Canadian-American Law Journal of the Gonzaga University School of Law, Spokane, WA.

2. Public Utilities Regulatory Policy Act of 1978, Pub. L. 95-617, 92 Stat. 3117 (codified at 16 U.S.C. §§ 2601-2645 and other scattered sections of U.S.C. (1982)).

3. *See, e.g.*, Energy Law Institute, *Federal Legal Obstacles and Incentives to the Development of the Small Scale Hydroelectric Potential of the Nineteen Northeastern United States* (1980); Scotch, "Small Hydropower Development and the Environment: A Survey of State and Federal Law," 5 Vt. L. Rev. 251 (1980); Burke, "Small Scale Hydroelectric Development and Federal Environmental Law: A Guide for the Private Developer," 9 B.C. Env. Aff. L. Rev. 815 (1981-1982).

4. 43 U.S.C. §§ 1301-1303, 1311-1315 (1982).

5. *Id.* § 1311(d).

6. 12 M.R.S.A. § 558 (1981).

7. 16 U.S.C. § 814 (1982).

8. 16 U.S.C. §§ 791a-793, 795-818, 820-825r (1982).

9. 16 U.S.C. § 797(e) (1982).

10. 16 U.S.C. § 803(a) (1982).

11. *See Udall v. Federal Power Commission*, 387 U.S. 428 (1967); *Pacific Power & Light v. Federal Power Commission*, 333 F.2d 689 (9th Cir. 1964), *cert. denied*, 379 U.S. 969 (1964). 16 U.S.C. § 811 also provides specifically that: "The Commission shall require the construction, maintenance, and operation by a licensee at its own expense of ... such fishways as may be prescribed by the Secretary of the Interior or the Secretary of Commerce as appropriate."

12. Arnold, "Emerging Possibilities for State Control of Hydro-electric Development," 13 Envtl L. Rep. (Envtl L. Inst.) 10135 (1983).

13. *Monongahela Power Co. v. Alexander*, 507 F.Supp. 385 (D.D.C. 1980) (decision pending D.C. Cir.).

14. *First Iowa Hydro-Electric Cooperative v. Federal Power Commission*, 328 U.S. 152 (1946); *see Town of Springfield v. Vermont Environmental Board*, 521 F. Supp. 243 (D.Vt.1981) (following *First Iowa*); *Town of Springfield v. McCarren*, 549 F. Supp. 1134 (D.Vt.1982) (following *First Iowa*), *aff'd* No. 82-7837 slip op. (2d Cir. Feb. 16, 1983), *cert. denied*, U.S. , 78 L. Ed. 2d 322 (1984).

15. *See* Arnold, *supra* note 32.

16. 33 U.S.C. § 401 (1982).

17. 33 U.S.C. § 403 (1982).

18. 33 C.F.R. § 320.3(f) (1982). *See Scenic Hudson Preservation Conference v. Callaway* 370 F.Supp. 162, 167 (S.D.N.Y. 1973), *aff'd per curiam*, 499 F.2d 127 (2d Cir. 1974) (holding that Corps § 10 authority is preempted by the licensing authority of the Federal Power Act).

19. For an exhaustive discussion of § 404, see, Blumm, "The Clean Water Act's Section 404 Permit Program Enters Its Adolescence: An Institutional and Programmatic Perspective," 8 Ecol. L.Q. 409 (1980); *also see* Parish & Morgan, "History, Practice and Emerging Problems of Wetlands Regulations: Reconsidering Section 404 of the Clean Water Act," 17 Land & Water L. Rev. 43 (1982).

20. "Fill" is very broadly defined in 18 C.F.R. § 323.2(n); "dams and dikes" are specifically included as activities that meet the definition, and therefore require § 404 permits.

21. *But see Monongahela Power Co. v. Alexander* 507 F. Supp. 385 (D.D.C. 1980) (decision pending, D.C. Cir.). This case holds that FERC has exclusive preemptive regulatory power over hydro-electric dams under the Federal Power Act, and therefore the Corps lacks jurisdiction to grant or deny 404 permits or, by implication, any other permits, such as § 10 permits. This decision directly conflicts with the holding in *Scenic Hudson Preservation Conference v. Calloway*, 370 F. Supp. 162 (S.D.N.Y. 1973), *aff'd per curiam*, 499 F.2d 127 (2d Cir. 1974). Although the D.C. District Court attempts to in effect "overrule" *Scenic Hudson*, conflict within the circuits can ultimately only be resolved by a Supreme Court decision or by clarifying Congressional legislation.

22. 33 U.S.C. § 1344(r) (1982).

23. 33 C.F.R. § 320.4(j) (1982).

24. 33 U.S.C. § 1341 (1982).

25. 33 C.F.R. § 325.2(b)(1)(2) (1982). See text *infra* at notes 105-112 for further discussion of Coastal Zone Management Act requirements.

26. 40 C.F.R. pt. 230 (1982).

27. 33 U.S.C. § 1344(b)(1) (1982).

28. 40 C.F.R. § 230.1 (1982).

29. 33 C.F.R. § 325.8(d)(6) (1982).

30. 33 U.S.C. § 1344(c) (1982).

31. As of the end of 1983, this authority had been invoked only once, in 1980, [14 Current Developments] Env't Rep. (BNA) 1395 (Dec. 2, 1983), but EPA has three section 404(c) veto proceedings currently pending. [15 Current Developments] Env't Rep. (BNA) 579 (Aug. 10, 1984).

32. Codified as amended at 42 U.S.C.A. §§ 4321-4370 (West 1977 & Supp. 1983).

33. *Id.* at § 4332.

34. 40 C.F.R. parts 1500-1508 (1982).

35. "Council on Environmental Quality Guidance on National Environmental Policy Act Regulations," 48 Fed. Reg. 34263 (July 28, 1983); *see* 40 C.F.R. § 1500.1 (1982).

36. *See, e.g.*, the statement of policy at 40 C.F.R. § 1500.1(d) (1982): "NEPA procedures must insure that environmental information is available to public officials and citizens before decisions are made and before actions are taken. The information must be of high quality. Accurate scientific analysis, expert agency comments and public scrutiny are essential to implementing NEPA."

37. 40 C.F.R. § 1500.14 (1982).

38. 40 C.F.R. § 1502.22 (1982).

39. 18 C.F.R. § 2.81(a)(1)(i) (1982).

40. For discussions of "worst case analysis," its uses and its deficiencies, *see* Comment, "CEQ's 'Worst Case Analysis' Rule for EISs: 'Reasonable' Speculation or Crystal Ball Inquiry?", 13 Envt'l L. Rep. (Envt'l L. Inst.) 10069 (1983); Comment, "Update: The NEPA Worst Case Analysis Regulation," 14 Envt'l L. Rep. (Envt'l L. Inst.) 10267 (1984).

41. Exec. Order No. 12,114, 44 Fed. Reg. 1,957 (1979) *reprinted in* 42 U.S.C. § 4321 (Supp. V 1981).

42. *Wilderness Society v. Morton*, 463 F.2d 1261 (D.C. Cir. 1972) (Court grants standing to Canadian Wildlife Federation and a Canadian citizen to intervene in NEPA suit against U.S. Secretary of Interior to require preparation of impact statement on Alaskan Pipeline; of particular interest is note 2 of that opinion, stating in part: "Interestingly, the pipeline appellee, but not the Government, asserts that claims of the kind asserted by appellants when made by non-United States citizens, are non-justiciable under the doctrine of separation of powers. We find the contention without merit ...."); *People of Enewetak v. Laird*, 353 F. Supp. 811 (D. Hawaii 1973) (determination that NEPA applies to U.S. Trust Territories); *Sierra Club v. U.S. Atomic Energy Commission*, No. 1867-73, 4 Envtl L. Rep. (Envtl L. Inst.) 20685 (D.D.C. Aug. 3, 1974) (Court orders AEC to prepare EIS dealing generally with the nuclear power export process, based on international and bilateral "Agreement/s for the Cooperation Concerning Civil Uses of Atomic Energy" concluded pursuant to 42 U.S.C. § 2153); *Environmental Defense Fund v. U.S. Agency for International Development*, No. 75 Civ. 500, 6 Envtl L. Rep. (Envtl. L. Inst.) 20121 (D.D.C. Dec. 5, 1975) (Court orders AID to promulgate NEPA regulations and to prepare programmatic EIS on all its activities supporting pesticide procurement and use abroad); *Sierra Club v. Coleman* (I), 405 F. Supp. 53 (D.D.C. 1975), *Sierra Club v. Coleman* (II), 421 F. Supp. 63 (D.D.C. 1976) (issuance of preliminary injunction against U.S. aid in construction of Darien Gap Highway in Panama and Columbia, pending preparation of Environmental Impact Statement and general compliance thereof and continuance of injunction following preparation of EIS based on inadequacy of EIS discussion of certain environmental impacts, primarily control of the spread of a cattle disease known as aftosa or "foot and mouth" disease), *rev'd sub nom. Sierra Club v. Adams*, 578 F.2d 389 (D.C. Cir. 1977) (questions of standing of plaintiff to challenge adequacy of statement regarding local effects discussed, assuming, without resolution of the question, the applicability of NEPA to construction in Panama, *Id.* at n. 14); *In the Matter of Babcock & Wilcox*, No. 50-571, 7 Envtl L. Rep. (Envtl L. Inst.) 30017 (N.R.C. June 27, 1977) (determination that West German citizens' group lacked standing to intervene in licensing proceeding for exporting nuclear reactor components to West Germany, and that a generic EIS on the nuclear export program having been prepared, no further EIS was required for actions affecting only the local environment of foreign nations); *Natural Resources Defense Council Inc. v. Eximbank*, No. 77-0080 (D.D.C. Jan. 14, 1979) (a catalyst for the issuance of Executive Order 12,114, case seeking to require Eximbank to comply with NEPA in its activities, resolved by Executive Order 12,114 § 3-4 excluding direct loans to finance imports from the general exemption from environmental assessment for export approvals, i.e., while a complete EIS need not be prepared, an environmental assessment must); *NORML v. U.S. Dept. of State*, 452 F.Supp. 1226 (D.D.C. 1978) (Court issues declaratory judgment that State Department must prepare EIS on impact *within United States* of U.S. sponsored pesticide spraying of marijuana and poppies in Mexico, without deciding the question of the extraterritorality of NEPA); *Natural Resources Defense Council Inc. v. Nuclear Regulatory Comm'n*, 647 F.2d 1345 (D.C. Cir. 1981) (Court decides that NRC export of nuclear reactor and related materials to the Philippines does not require environmental assessment either under the Atomic Energy Act or NEPA, leaving open the question of whether NEPA might be applicable to some other kind of major federal action abroad).

43. *See, e.g.*, Note, "Environmental Restrictions: Extraterritorial Reach of United States Environmental Quality Standards - *Natural Resources Defense Council, Inc. v. Nuclear Regulatory Comm'n*, 647 F.2d 1345 (D.C. Cir. 1981)," 2 Harv. Int'l L.J. 676 (1981); Comment, "Renewed Controversy over the International Reach of NEPA," 7 Envtl. L. Rep. (Envtl L. Inst.) 10204 (1977); Note, "The Extraterritorial Scope of NEPA's Environmental Impact Statement," 74 Mich. L. Rev. 349 (1975); Robinson, "Extraterritorial En-

vironmental Protection Obligations of Foreign Affairs Agencies: The Unfulfilled Mandate of NEPA," 7 N.Y.U. J. of Int'l L. and Pol. 257 (1974).

44. *See* "CEQ Draft Regulations on Applying NEPA to Significant Foreign Environmental Effects," [8 Current Developments] Env't. Rep. (BNA) 1495 (1978).

45. *E.g.*, Note, "Agency Responses to Executive Order 12,114: A Comparison and Implications," 14 Cornell Int'l L.J. 481 (1981); Note, "The Extraterritorial Application of NEPA under Executive Order 12,114," 13 Vand. J. Transnat'l L. 173 (1980); Comment, "Federal Agency Responsibility to Assess Extraterritorial Impacts," 14 Tex. Int'l L.J. 425 (1979); Comment, "President Orders Environmental Review of International Actions," 9 Envtl. L. Rep. (Envtl. L. Inst.) 10011 (1979).

46. Exec. Order No. 12,114, § 1-1, *supra* note 41.

47. 33 U.S.C. §§ 1401-1445 (1982)

48. 33 U.S.C. §§ 1501-1524 (1982).

49. Exec. Order No. 12,114, § 1-1, *supra* note 41.

50. *Id.* § 2.

51. *Id.* § 2-4(a)(iii)

52. *Id.* § 4(b)(i).

53. *Id.* § 2-5

54. *Id.* § 3-1.

55. *Id.* § 3-5 (emphasis added).

56. *But see* Comment, 14 Tex. Int'l L.J. 425, *supra* note 45, at 450; Comment, 9 Envtl. L. Rep. 10,011, *supra* note 45, at 10,015.

57. *See* Exec. Order No. 12,114 § 2-3(b), *supra* note 41.

58. *See* discussion *supra* note 42.

59. Comment, 9 Envtl. L. Rep. 10,011, *supra* note 45, at 10,016.

60. Comment, 14 Tex. Int'l L.J. 425 *supra* note 45, at 450-51.

61. Codified as amended at 16 U.S.C. §§ 661 to 669g-l (1982).

62. *Id.* § 662(a).

63. *Id.* §§ 662(a),(b).

64. *Cf.* 42 U.S.C. § 4332 and 16 U.S.C. § 662(a).

65. The reporting agency is defined earlier in the section as "any agency of the federal government responsible for engineering surveys and construction of such projects when such reports are presented to the Congress or to any agency or person having the authority ... to authorize the construction of water resource development

projects...." This would apply to the Army Corps of Engineers, giving them rather than U.S.F.W.S. the determination of which measures are "justifiable".

66. 16 U.S.C. § 662(b) (1982) (emphasis added).

67. *See* U.S. Fish and Wildlife Service Mitigation Policy, 46 Fed. Reg. 7644, 7647, 7654 (Jan. 23, 1981), interpreting Fish and Wildlife's authority to be advisory only, with the permitting agency having the final say. However, 16 U.S.C. § 663(a) provides that when the United States (as opposed to a utility or private developer) undertakes development of waterways, "adequate provision ... shall be made for the conservation, maintenance and management of wildlife resources ... and ... habitat, including the development and improvement of such wildlife resources...." Section 18 of the Federal Power Act, 16 U.S.C. § 811 requires the inclusion of fishways when recommended (*see* note 31, *supra*), but this has been interpreted by the Department of Interior's Fish and Wildlife Service to apply only to hydro-electric projects in "specific withdrawn public lands." U.S. Fish and Wildlife Service Mitigation Policy, *supra*, at 7660. Finally, § 30 of the Public Utilities Regulatory Policy Act of 1978 (PURPA), 16 U.S.C. § 823a provides that for certain conduit hydroelectric facilities to qualify for an exemption from licensing they must comply with terms and conditions enforced by U.S.F.W.S. and/or the responsible state agency.

68. Pub. L. No. 93-205, 87 Stat. 884 (codified as amended at 16 U.S.C. §§ 1531-1543 (1982)).

69. Pub. L. No. 92-522, 86 Stat. 1027 (codified as amended at 16 U.S.C. §§ 1361-1407 (1982)).

70. *Id.* § 1361(b).

71. *See, e.g., Pittston Co. v. Endangered Species Committee*, 14 (Env't Rep. Cas.) 1257 (D.D.C. 1980); *Roosevelt Campobello International Park v. United States Environmental Protection Agency*, 684 F.2d 1041 (1st Cir. 1982).

72. 16 U.S.C. § 1536 (1982).

73. *Id.* Sub-§(c).

74. *TVA v. Hall*, 437 U.S. 153 (1978).

75. 16 U.S.C. § 1536(h) (1982).

76. *See Conservation Law Foundation v. Andrus*, 623 F.2d 712 (1st Cir. 1979) and *North Slope Borough v. Andrus*, 642 F.2d 489 (D.C. Cir. 1980) (addressing outer continental shelf oil and gas leases and endangered whales off Georges Bank and in the Beaufort Sea, respectively.)

77. 16 U.S.C. §§ 470 to 470w-6 (1982).

78. *Id.* § 470a(a)(1).

79. 16 U.S.C. §§ 469 to 469c-l (1982).

80. *Id.* at § 469

81. 16 U.S.C. §§ 1451-1464 (1982).

82. Pub. L. No. 94-370, 90 Stat. 1013.

83. 16 U.S.C. § 1456a(c)(3) (1982).

84. *See* P. Eliopoulos, *Coastal Zone Management: Program at a Crossroads* (Env't Rep. (BNA) Monograph No. 30, (1982)).

85. Maine's plan, approved in September, 1978, is composed of eleven core laws including the Alteration of Coastal Wetlands law, 38 M.R.S.A. §§ 471-476. This statute requires that all applicants for a permit demonstrate to the municipality or the State Board of Environmental Protection, whichever is administering the law, that the proposed activity will not

> unreasonably interfere with existing recreational and navigational uses, nor cause unreasonable soil erosion, nor unreasonably interfere with the natural flow of any waters; nor unreasonably harm wildlife or freshwater, estuarine or marine fisheries; nor lower the quality of any waters.

Id. § 474. What is "unreasonable" is a matter of agency discretion.

86. 16 U.S.C. § 1456(c)(3)(a) (1982).

87. *Id.*

88. *Id.*

89. 328 U.S. 152 (1946).

90. Passamaquoddy Tribal Council, *Half Moon Cove Tidal Power U.S.A.: Application for License for Major Unconstructed Project for Project No. 3035 Before the Federal Energy Regulatory Commission*, ii-v (June 16, 1983).

91. 38 M.R.S.A. §§ 630-636 (West Supp. 1979-1983).

92. *Id.* § 634(1)

93. 18 C.F.R. § 4.41(f)(2)(vii)(A) (1982).

94. P.L. 1983, ch. 458 (codified at 12 M.R.S.A. §§ 401-406 and scattered sections, 30 M.R.S.A. §§ 1961-1969 and scattered sections, 33 M.R.S.A. § 668 and 38 M.R.S.A. §§ 630-636 (Supps. 1983)).

95. Letter from C.M. Butler, Chairman, FERC to Gov. Joseph Brennan (Aug. 3, 1982), cited in Arnold, *supra* note 12, at 10,140 n. 69.

96. *Supra* note 12.

97. S. 2361, 98th Cong., 2d Sess. (1984).

98. 507 F.Supp. 385 (D.D.C. 1980) (decision pending D.C. Cir.). In another case concerning preemption of federal authority, *Escondido Mutual Water Co. v. LaJolla, Rincon, San Pasqual and Pala Bands of Mission Indians*, 52 U.S.L.W. 4588 (U.S. May 15, 1984) the United States Supreme Court recently held that under the terms of the Federal Power Act, FERC had to impose on the license those conditions that the Secretary of Interior deemed "necessary for the protection and utilization of the reservation." However, this decision does not necessarily preclude the District of Columbia Circuit from affirming the holding in *Monongahela* that Corps regulation under § 404 is preempted; the Federal Power Act nowhere specifically refers to conditions of the Corps with respect to wetlands being included in licenses, but then the Corps has only in the past decade undertaken wetlands regulation. The scope of FPA preemption remains uncertain except as to the issue specifically decided in *Escondido*.

99. *See generally* Randle, "Coastal Energy Siting Dilemmas," 21 Nat. Res. J. 125 (1981)

100. *See* letter of transmittal from Ralph Dana, Governor, Passamaquoddy Tribe at the Pleasant Point Reservation, to Kenneth F. Plumb, Secretary, FERC (June 16, 1983), *reprinted in* Passamaquoddy Tribal Coun-

cil, *supra* note 114 (indicating that the license application represented seven years of preparation). FERC rejected the revised license application for deficiencies on April 20, 1984; thus the tribal council lost its FERC preliminary permit giving it priority for development of the Half Moon Cove site.

101. Federal Power Act § 205, 16 U.S.C. § 824d (1982).

102. Federal Energy Regulatory Commission, *1979 Annual Report* 42.

103. *See* Federal Power Act §§ 205, 206, 16 U.S.C. §§ 824d, 824e (1982).

104. Federal Power Act § 206, 16 U.S.C. § 824e.

105. Section 102 of the National Environmental Policy Act of 1969 (NEPA), 42 U.S.C. § 4332, requires Federal agencies to prepare detailed reports on the environmental effects of proposed major actions. *See* 40 C.F.R. § 1501.4 and pt. 1502 (1982) (Council on Environmental Quality regulations regarding when and how to prepare an "environmental impact statement.") *Also see* 18 C.F.R. §§ 2.80-81 (1983) (FERC regulations implementing NEPA.)

106. 42 U.S.C.A. §§ 4321-4370 (West 1976 & Supp. V 1981).

107. Federal Power Act, § 202(e), 16 U.S.C. § 824a(e) (1982). The Secretary of the Department of Energy has succeeded to the authority under this provision formerly vested in the Federal Power Commission. Department of Energy Organization Act, Pub. L. 95-91, §§ 301(b), 402(f) 91 Stat. 577, 583, (codified at 42 U.S.C. §§ 7151(b) and 7172(f) (Supp. V 1982)).

108. Federal Power Act §202(f), 16 U.S.C. § 824a(f) (1982), provides, *inter alia* that:

> The ownership or operation of facilities for transmission or sale at wholesale of electric energy which is ...(b) generated in a foreign country and transmitted across an international boundary into a State and not thereafter transmitted into any other State, shall not make a person a public utility subject to regulation under other provisions of this subchapter. The State within which any such facilities are located may regulate any such transaction insofar as such State regulation does not conflict with the exercise of the Commission's powers under or relating to subsection (e) of this section.

It is likely that the owner of the transmission facilities for receiving Nova Scotian Tidal Power would be a public utility otherwise subject to FERC jurisdiction. Recent proposals by the Central Maine Power Company to reorganize and to set up nonregulated affiliates have met with State opposition in the form of intervention by the Maine Public Utilities Commission before the Securities and Exchange Commission and the passage of legislation requiring PUC approval of any utility reorganization. P.L. 1981, ch. 672 (codified at 35 M.R.S.A. § 104 (3-A) (Supp. 1978-1983)).

109. 3 C.F.R. 970 (1949-1953 Comp.), as amended by Exec. Order No. 12,038, 3 C.F.R. 136 (1978), *reprinted in* 15 U.S.C. § 717b note (1982).

110. *Id.* § 1(a)(3).

111. *Id.* § 1(b).

112. *Id.* § 1(a)(3), which provides: "The Secretary of Energy shall have the power to attach to the issuance of the permit and to the exercise of the rights granted thereunder such conditions as the public interest may in its judgment require."

113. DOE Delegation Order (D.O.) No. 0204-4, para. 7 (1977) *reprinted at* 10 C.F.R. § 1001.1, app. (1983). This Order provides

> There is hereby delegated to the Administrator of ERA the authority to adopt rules, issue orders, licenses and allocations, collect fees and take such other action as may be necessary and appropriate to

administer the following functions:

...

7. The exportation and importation of electric energy pursuant to the provisions of Section 202(e) of the Federal Power Act ... as amended, and Executive Order No. 10485.

10 C.F.R. § 205.320(a) (1983) further delegates this authority to ERA's Office of Utility Systems.

114. 10 C.F.R. §§ 205.320-327.

115. *Id.* at § 205.321.

116. Telephone conversation with Lise Howe, Office of General Counsel, DOE (April 22, 1983).

117. 10 C.F.R. § 1021.2 (1983) (adopting CEQ guidelines codified at 40 C.F.R. §§ 1500-1508). For general discussion of NEPA see text *supra* at notes 58-64.

118. 10 C.F.R. § 205.322(c)(1983); *But see* 40 C.F.R. § 1502.14 (requiring rigorous exploration of all alternatives to a project including even alternatives not within the jurisdiction of the lead agency preparing the EIS) and 18 C.F.R. pt. 2, app. A § 8.2 (consideration of energy alternatives in the preparation of EIS's under the Federal Power Act).

119. U.S. DOE, *EIS: 300 kV International Submarine Transmission Line - Erie, Pennsylvania to Nanticoke, Ontario, Canada, General Public Utilities Corporation*, pt. 2 (1982) and U.S. DOE, *EIS: New England/Hydro-Quebec 450 kV Direct Current Transmission Line Interconnection* pt, 2 (1984). (discussing such alternatives to the Canadian transmission line interconnections as purchasing power from other U.S. utilities or building other generating capacity.)

120. Tidal Power Corporation, *Fundy Tidal Power Update '82*, 5, 24 (1982).

121. *Id.* at 6.

122. *Id.* at 14-15. *See* Quick, "Canadian Power and the New England Energy Crisis," 6 New Eng. J. Bus & Econ. 55 n.1, and 63 (1980). (Mr. Quick suggests that the one major New England-Canadian transmission link, a 345 kV line between Maine and New Brunswick, has "inadequate capacity to handle major interchanges of electricity on the scale envisioned." The discussion is largely in the context of Quebec hydropower imports.) *See also*, "Report on United States/Canada Electric Power Exchange Study: New England New Brunswick," *reprinted in* U.S.D.O.E. and Canadian Department of Energy, Mines and Resources (D.E.M.R.), *United States/Canada Electricity Exchanges* app. D (Feb. 1979). ("It should be noted that in addition to the direct 'across the border' transmission interconnection facilities which would be required to permit substantial increases in power transfers, significant internal transmission reinforcement would also be necessary." *Id.* at D-11.)

123. Quick, *supra* note 122, at 64.

124. 10 C.F.R. § 205.323(e) (1983).

125. Some earlier tidal power development options included the construction of pumped storage hydro facilities in New England to absorb non-peak capacity excess. *See United States/Canada Electricity Exchanges*, *supra* note 122, at 62. Such a major construction project in the U.S., requiring FERC and other federal permits, would provide an even stronger base for a total project investigation--including the effects from Canadian barrage construction. However, the latest project designs do not contemplate the use of pumped storage. TPC, *Fundy Tidal Power Update '82*, *supra* note 120, at 2-3.

126. 528 F.2d 38 (2d Cir. 1975).

127. *Id.* at 40. The court found that it lacked jurisdiction and did not reach the merits of the case.

128. 15 U.S.C. § 793(d) (1982). This provision was part of the Energy Supply and Environmental Coordination Act of 1974, Pub. L. 93-319, 88 Stat. 246, which was passed in the face of an OPEC oil embargo and the accompanying American "energy crisis." Specifically it provides:

> In order to expedite the prompt construction of facilities for the importation of hydro-electric energy thereby helping to reduce the shortage of petroleum products in the United States, the Federal Power Commission is hereby authorized and directed to issue a Presidential permit pursuant to Executive Order 10485 of September 3, 1953, for the construction, operation, maintenance and connection of facilities for the transmission of electric energy at the borders of the United States *without preparing an environmental impact statement pursuant to section 102 of the National Environmental Policy Act of 1969* ... for facilities for transmission of electric energy between Canada and the United States in the vicinity of Fort Covington, New York.

(Emphasis added.)

129. *See supra* note 112 and accompanying text.

130. 16 U.S.C. §§ 1451-1464 (1982). But the State does not have the final say; the Secretary of Commerce can overrule a state's denial of consistency. *See supra*, notes 87-89 and accompanying text.

131. 38 M.R.S.A. §§ 481-490 (1978 & Supp. 1979-83).

132. *See infra* notes 149-158 and accompanying text.

133. Entry of Nova Scotia Tidal Power into United States power grids other than through Maine appears unlikely, primarily for geographical reasons. (There could be additional political and technical problems, not to mention economic questions, with any other route, e.g., wheeling through Quebec to the proposed Quebec-Vermont link. *See* Quick, *supra* note 146 at 62, re asynchronism of Quebec with the New England Power Pool (NEPOOL).)

134. 35 M.R.S.A. § 13-A (Supp. 1983-84).

135. 35 M.R.S.A. § 13-B (Supp. 1983-84).

136. 35 M.R.S.A. §§ 13-A and 13-B (Supp. 1983-84).

137. TPC, *Fundy Tidal Power Update '82*, *supra* note 120, at 27. The update does recognize that utility regulatory policy in the United States may preclude use of the joint venture structure if construction work in progress (CWIP) cannot be used in the rate base, but states that "[t]hese considerations are not reflected in the projections made during the course of this study, but they do have a bearing on the feasibility of the joint venture approach." *Id.* at 29.

138. 314 A.2d 800 (Me. 1974).

139. *Id.* at 809-811. *See* 35 M.R.S.A. § 2306, providing, *inter alia*, that in eminent domain proceedings: "Environmental factors to be considered for proper location of a transmission line or gas pipeline shall not be subject to review by the Public Utilities Commission when the location of the transmission line has received site location of development approval under Title 38, section 484." This provision was added by P.L. 1977, ch. 374, § 2, apparently in response to the *Bangor Hydro-Electric* decision which, while precluding consideration of environmental factors by the PUC in the approval of 100 kV plus transmission lines pursuant to 35 M.R.S.A. § 13-A, found PUC consideration of the environmental effects mandated for lower-powered lines in eminent domain proceedings. A site location of development permit is required prior to the acquisition of land by eminent domain. 38 M.R.S.A. § 484 (1978).

140. 35 M.R.S.A. § 13-A (Supp. 1983-84).

141. *Id.*

142. *See* Energy Information Administration (EIA), *U.S.-Canadian Electricity Trade*, 40 (1982) ("Property owners object to transmission lines both because of the loss of land for the right-of-way, and because of fears about the safety of high voltage power lines. The objections to high voltage power lines have led to extensive regulatory proceedings, appeals, and litigation significantly delaying construction of such facilities.") Opposition to lines could be mitigated, however, if the new lines parallelled existing transmission lines, using the existing right-of-way.

143. Quick, *supra* note 122, at 62-63.

144. *See Effects of Proposed Hydro-Electric Project in the Bay of Fundy: Hearing Before the Senate Commission Env't and Public Works*, (hereinafter cited as Senate Tidal Power Hearing), 98th Cong., 1st Sess. 74 (1983) (post-hearing submission of G.C. Baker, Exec. Vice-President, Tidal Power Corporation).

145. 35 M.R.S.A. § 2483 (1978).

146. The Shoreland Zoning Act, 12 M.R.S.A. §§ 4811-4817 (1978 & Supp. 1983-84), sets guidelines for municipal shoreland zoning and imposes regulations in the event a locality fails to adopt its own shoreland zoning ordinance.

147. 12 M.R.S.A. §§ 681-689 (1978 & Supp. 1983-84). "Unorganized and deorganized areas ... include all areas located within the jurisdiction of the State of Maine, except areas located within organized cities and towns, and Indian reservations." *Id.* at § 682(1). These areas, lacking municipal government and having few inhabitants, are administered by LURC in the public interest.

148. For a summary review of the Maine statutes applicable to transmission line construction, see "Report on United States/Canada Electric Power Exchange Study: New England New Brunswick," *supra* note 122, at D-15 to D-17.

149. 38 M.R.S.A. § 481 (Supp. 1979-83).

150. 38 M.R.S.A. § 482(2) (Supp. 1979-83).

151. 38 M.R.S.A. § 484 (1978 & Supp. 1979-83).

152. Maine Department of Environmental Protection Regulations, ch. 375 (1979).

153. *Id.*, ch. 372, § 1 (1979)(emphasis added).

154. *See In re Ryerson Hill Solid Waste Disposal Site*, 379 A.2d 384 (Me. 1977); *In re Maine Clean Fuels*, 310 A.2d 736, 742 (Me. 1973) (quoting *In re Spring Valley Development*, 300 A.2d 736, 753-54 (Me. 1973) for the proposition that the Site Location law "did not grant the commission the authority to determine where the location of a development must be but rather it gave the Commission authority to measure the proposal and location against statutory standards and to apply reasonable terms and conditions...").

155. *See supra* note 151 and accompanying text.

156. Assuming that any new transmission line would be run parallel to the existing 345 kV line from Keswick, New Brunswick to Orrington, Maine, it is worth noting that the area is sparsely inhabited.

157. *See* Natural Resources Council of Maine "Tidal Power", *Maine Environment* (June-July 1982); Senate Tidal Power Hearing, *supra* note 144, at 50, 265 (statements of Mark Ishkanian, Natural Resources Council of Maine).

158. "Report on United States/Canada Electric Power Exchange Study: New England-New Brunswick," *supra* note 122 at D-18.

159. *See* discussion *supra* at notes 41-60.

160. *See generally* Taylor, "The International Joint Commission and Bay of Fundy-Gulf of Maine Region Transboundary Environmental Impacts" (Marine Law Institute, Donner Project Working Paper -6, 1984) (unpublished).

161. *See* Taylor, "Liability for Transboundary Pollution and State Responsibility," 66 (Marine Law Institute, Donner Project Working Paper -4, 1984) (unpublished) and sources cited therein. Private plaintiffs seeking remedies for transboundary damage could also be met with the defense of statutory or administrative authority. W. Rodgers, *Environmental Law* 3136-42 (1977); R. Franson & A. Lucas, *Canadian Environmental Law* Vol. 1, 3.1.2.2; *e.g.*, *New England Legal Foundation v. Costle*, 666 F.2d 30, 32-33 (2d Cir. 1981). This defense has been weakened by both Canadian and United States Courts in recent years. Taylor, "Private Remedies for Transboundary Environmental Injury Due to Coastal Energy Development in the Bay of Fundy-Gulf of Maine Region," 17 (Marine Law Institute, Donner Project Working Paper -5, 1984) (unpublished). *See generally id.* for a discussion of the problems facing private plaintiffs seeking redress for transboundary injury in this region, e.g., possibility that United States private plaintiffs would have no standing to sue for property damage in Canadian courts.

162. Senate Tidal Power Hearing, *supra* note 144, at 9 (statement of G.C. Baker, Exec. Vice-President, Tidal Power Corporation).

163. *Id.* at 69, 74-75 (statement and post-hearing submission of G.C. Baker).

164. Similar arguments might be made with regard to construction of other pollution sources such as factories for manufacture of goods for export. Indeed, the President of the North American Electric Reliability Council (NERC) recently noted: "When one considers a Canadian [electricity] import as being the same as any other international trade transaction, it appears reasonable that the Federal Government should have some control." Letter from Walter D. Brown, President, NERC, to Dexter J. Peach, Director, Energy and Minerals Division, United States General Accounting Office, (June 15, 1982) *reprinted in* GAO, *Report to the Congress: Clear Federal Policy Guidelines Needed for Future Canadian Power Imports*, app. VIII (1982). However, there are numerous reasons for asserting the uniqueness of the energy export-import situation. For example, as one recent report notes: "Trade in electricity is isolated from other trade flows, because electricity can move only along specially constructed transmission lines." Energy Information Administration, *U.S.-Canadian Electricity Trade* 21 (November, 1982). Electricity is, in any event, an industry heavily regulated in the public interest. Expansion of this public interest review in the case of power imports would be a natural adjunct of existing review; to further quote the President of NERC: "The utility systems, and possibly their respective reliability councils, are quite capable of handling these matters [i.e., electricity imports and interconnections] themselves without any need for intervention from the Federal government, *other than in their mandated lead role in environmental matters*." (Emphasis added.) Letter from Brown to Peach, *supra*. In other words, the role of electric utilities is not to examine environmental impacts of their actions--unless it is federally mandated.

165. In the alternative, the President could perhaps amend Executive Order 10,485 to require that environmental impact assessments for transboundary transmission lines cover transboundary effects of the power plant located abroad and that upgrading of lines be subject to NEPA if undertaken for the importation of power from a plant that might have significant effects. But the real time to investigate these issues is when the contracts are signed before transmission line construction.

166. For a general discussion of such constitutional bases of power, *see* VanderZwaag, *Fundamentals of Fundy Tidal Power: The Canadian Decision-Making Framework and Decision-Making Options*, 3 Can.-Amer. L.J. 5-16 (1984).

167. For a detailed discussion of the new constitutional amendment and the definitional uncertainties raised by the words "conservation and management," *see* W. Moull, "The Legal Effect of the Resource Amendment - What's New in Section 92A?" in J. Meekison, R. Romanow, W. Moull, *Origins and Meaning of Section 92A: The 1982 Constitutional Amendment on Resources* (1985), pp. 50-53.

168. *Canada - Nova Scotia Agreement on Offshore Oil and Gas Resource Management and Revenue Sharing* (March 2, 1982). For a discussion of the agreement's implementation, *see* Doucet, *Canada-Nova Scotia Offshore Agreement: One Year Later*, 22 Alta. L. Rev. 132 (1984). The agreement is likely to be renegotiated in 1986 in order to provide Nova Scotia with a more favorable revenue-sharing and management arrangement along the lines of the offshore accord reached between Newfoundland and the federal government. *The Chronicle-Herald* (Halifax) (January 3, 1986), pp. 1, 20.

169. For a discussion of the paramountcy doctrine in relation to Section 92A, *see* W. Moull, *supra* note 167, pp. 53-54.

170. Environmental Assessment and Review Process Guidelines Order, SOR/84-467.

171. Crown corporations listed in Schedule D to the *Financial Administration Act* would be subject to the Process if it is corporate policy to apply the Process and if the application of the Process is within the legislative authority of the corporation. SOR/84-467, s. 7. Such corporations include: Air Canada, Canada Deposit Insurance Corp., Canadian Broadcasting Corp., Cape Breton Development Corp., Central Mortgage and Housing Corp., Eldorado Aviation Ltd., Eldorado Nuclear Ltd., Export Development Corp., National Railways, Northern Transportaton Co. Ltd., Polymer Corp. Ltd., The St. Lawrence Seaway Authority, the Seaway International Bridge Corp. Ltd., Canadian Development Investment Corp., VIA Rail Canada, Federal Business Development Bank, Petro-Canada, and Teleglobe Canada. *Financial Administration Act*, R.S.C. 1970, C. F-10, Schedule D, *as amended by* S.C. 1974-75-76, 14, s. 56; C. 61, s. 15; C. 77, s. 8; SOR/78-287 and SOR/82-854.

172. R.S.C. 1970, C. F-14.

173. R.S.C. 1970, C. N-19.

174. Tidal Power Corporation Act, S.N.S. 1970-71, C. 21, ss. 9, 12, 14, 15.

175. R.S.N.S. 1967, C. 335. The permitting requirements of the Environmental Protection Act might also apply through broad readings of the words "contaminant" and "detrimental variation or alteration" and "pollution" defined by the Act. A permit issued pursuant to the Water Act would be deemed to be granted under the Environmental Protection Act. Environmental Protection Act, S.N.S. 1973, C. 6, s. 36.

176. Lands and Forests Act, R.S.N.S. 1967, C. 163, ss. 33, 38.

177. Electric Power Act, R.S.N.B. 1973, C. E-5.

178. R.S.N.B. 1973, C. C-6.

179. *Ibid.*, s. 9, *as amended by* S.N.B. 1975, C. 12, s. 4.

180. *Ibid.*, s. 32(1), *as amended by* S.N.B. 1975, C. 12, s. 8.

181. S.N.B. 1980, C. 38.1.

182. R.S.C. 1970, C. N-6.

183. *Ibid.*, s. 46(1) *as amended by* S.C. 1980-81-82, C. 115, s. 14.

184. *Ibid.*, s. 82(1) *as amended by* S.C. 1980-81-82, C. 115, s. 24.

185. National Energy Board Rules of Practice and Procedure, C.R.C. 1978, C. 1057, s. 3(2).

186. *National Energy Board Report to the Governor in Council in the Matter of the New Brunswick Electric Power Commission* 33 (July 1972).

187. National Energy Board Part VI Regulations, C.R.C. 1978, C. 1056, s. 6(2) (aa).

188. National Energy Board Rules of Practice and Procedure, C.R.C. 1978, C. 1057, Part III(11).

189. National Energy Board, *Environmental Guidelines for International Power Lines* (March 1981).

190. *National Energy Board Reasons for Decision in the Matter of Tenneco LNG Inc., Canadian Lowell Gas Ltd., Lorneterm LNG Limited and Trans Canada Pipelines (New Brunswick) Limited* (November 1977).

191. *National Energy Board Reasons for Decision in the Matter of Trans Mountain Pipe Line Company Ltd. and Foothills Oil Pipeline Company Ltd.* (January 1980).

192. *National Energy Board Reasons for Decision in the Matter of Trans Mountain Pipe Line Company Ltd.* (May 1981).

193. S.C. 1978-79, C. 13.

194. Environmental Assessment and Review Process Guidelines Order, SOR/84-467, ss. 27(2), 29(2).

195. *Ibid.*, s. 35(d).

196. Federal Environmental Assessment Review Office, *Environmental Assessment Panels: Procedures and Rules for Public Meetings* (1985).

197. *Environmenal Assessment and Review Process Guidelines Order*, SOR/84-467, s. 33(1)(d).

198. R.S.C. 1970, C.N-19, s. 5(1) (emphasis added).

199. Watercourse Alteration Regulation - Clean Environment Act, Reg. 82-233, s. 3(3).

200. *Ibid.*, s. 12(1).

201. *Ibid.*, s. 9.

202. *Committee for Justice and Liberty Foundation v. Interprovincial Pipeline (NS) Ltd.*, [1982] 1 F.C. 619, 11 C.E.L.R. 97 (F.C.A.).

203. *Essa v. Atomic Energy Control Board* (1981), 10 C.E.L.R. 142, 143 (F.C.T.D.) *reversed on other grounds sub. nom, Tosorontio v. Atomic Energy Control Board* (1981), 10 C.E.L.R. 145 (F.C.A.D.).

204. R.S.C. 1970, C. 10 (2nd Supp.).

205. *Ibid.*, s. 28. For debate as to how far courts should engage in substantive versus procedural review, *see* Mullan, *Natural Justice and Fairness - Substantive as well as Procedural Standards for the Review of Administrative Decision-Making?*, 27 McGill L.J. 250 (1982) and Grey, *Can Fairness be Effective?*, 27 McGill L.J. 360 (1982).

206. SOR/84-467.

207. *Ibid.*, s. 20. Public review would also be required when potential adverse environmental effects are significant as determined according to criteria cooperatively developed by FEARO and the initiating department.

208. For a discussion of the standing requirement for judicial review in Canada and the possible need for a private individual to demonstrate something more than a violation of a public right, *see* D. Scott, "Judicial Review in the Federal Court of Canada," in the Law Society of Upper Canada Bar Admission Course Materials, *Charter of Rights and Administrative Law* (1983), pp. 84-89.

209. Section 18 of the Federal Court Act provides a general right to seek relief (injunction, *certiorari*, prohibition, *mandamus*, *quo warranto*, or declaratory) from the Trial Division of the Federal Court against any federal board, commission or other tribunal.

210. For a discussion of the less onerous procedural requirements generally imposed on investigative tribunals without final decisionmaking powers, *see* Mercer, *Recent Developments in Canadian Law: Administrative Law II*, 16 Ottawa L. Rev. 645, 673 (1984).

211. *See e.g.*, *Attorney General of Canada v. Inuit Tapirisat of Canada*, [1980] 2 S.C.R. 735, where the Supreme Court of Canada refused to impose a duty of fairness on federal Cabinet review of a rate increase decision by the Canadian Radio-Television and Communications Commission because of the essentially legislative function exercised by the Cabinet, the political nature of the Cabinet's decision, and the practical difficulties raised by imposing procedural protections to Cabinet decisionmaking. For a general discussion of judicial reluctance to supervise legislative functions and ministerial discretion, *see* J. Cowan, "Recent Developments in Judicial Review" in the Law Society of Upper Canada Bar Admission Court Materials, *Charter of Rights and Administrative Law* (1983), pp. 15-16. 30-47.

212. For a detailed discussion of the creative possibilities the Charter might hold for the development of environmental law, *see* Stevenson, *A New Perspective on Environmental Rights After the Charter*, 21 Osgoode Hall L.J. 390 (1983).

Chapter 4

Effectiveness of Canadian and U.S. Environmental Assessment Processes in Evaluating Transboundary Impacts

Judith Spiller and John Roanowicz

Introduction

Environmental assessment serves several functions. First, it establishes a factual base for use in decisionmaking. Through participation of scientific and technical experts as well as the public, it gathers in one place a wide range of information and a diversity of opinion concerning a proposed project's merits. The process also opens decisionmaking to public scrutiny and at its best builds public confidence in government institutions. Finally, environmental assessment seeks to be comprehensive, avoiding the narrow decisionmaking often encountered with statutes governing single issues or activities.

The assessment process works most effectively in addressing a proposed activity's direct environmental effects. Direct effects are those which arise in a geographic and temporal proximity to the project and thus may directly be attributed to it. Environmental assessment, however, encounters difficulties when called upon to evaluate indirect impacts. This category includes distant, low-level, cumulative, or long-term effects. Indirect effects are often difficult to identify, to quantify, and to attribute to a source.

As a consequence, indirect effects create uncertainty in defining the nature and boundaries of possible impacts and in making predictions about the likelihood of those impacts. This uncertainty often sparks debate among experts. Thus, evaluators are required to reach a decision about a project's merits based on limited information about which there may be little expert agreement. In practice, they may ignore this class of impacts or through oversimplification misrepresent them.

Difficulty in characterizing indirect impacts becomes magnified in dynamic and complex ecosystems like that of the Bay of Fundy-Gulf of Maine. Here, large-scale ocean circulation and extreme tidal ranges dominate biological interactions. The factors driving the circulation - astronomical forces, incoming solar radiation, weather patterns, basin geometry, and the forcing effects of the Atlantic Ocean - exhibit significant variation and may act to magnify or block each other in ways which are poorly understood. These physical factors also influence the abundance and location of the biota. While a general understanding of the physical dynamics and their effects on biological parameters exists, this understanding proves of little use in projecting the location and extent of specific events.

These problems are compounded when jurisdictional boundaries are present. These boundaries seldom coincide with the ecosystem's physical and biological properties. The Canadian-U.S. boundary established by the World Court in 1984 roughly bisects the Bay of Fundy-Gulf of Maine basin (Figure 9). Water moves in a counterclockwise direction along the coast of western Nova Scotia around the Bay of Fundy and out along New Brunswick to the coast of Maine where it swings down past New Hampshire and back to Nova Scotia. The biota, including many important commercial fin- and shellfish species, travel in this current or rely on it to supply food sources. Species may spawn on the Canadian boundary but breed on the U.S. side, or they may be caught as juveniles on the U.S. side and as adults in the Canadian waters. Thus, changes in the environment on one side of the boundary may affect the environment and species dependent upon it on the other side. Similarly, human activities on one side, fishing for example, influence resources on the other side of the boundary.

Environmental assessment, when applied to a region like the Bay of Fundy - Gulf of Maine, should account both for direct and indirect impacts. The presence of an international boundary, however, may introduce assessment processes with differing goals and procedures. Ideally, assessment on either side of the boundary should be complementary, each placing the same weight on the goal of environmental protection while utilizing resources for the benefit of society.

Evaluating the effectiveness of environmental assessment in a transboundary environment raises two central questions. First, to what extent do present assessment procedures require consideration of indirect effects and transboundary inpacts? Second, how have the Canadian and American processes actually dealt with indirect effects in the transboundary context? This chapter will explore these questions, first by comparing the relevant requirements of the Canadian environmental assessment and review process (EARP) and the U.S.'s process under the National Environmental Policy Act (NEPA). It then examines the application of both procedures where indirect effects have occurred across the Canadian-U.S. boundary.

Environmental Assessment Processes in Canada and the United States: Requirements to Consider Indirect and Transboundary Effects

Canadian and Provincial Environmental Assessment Processes

At the provincial decisionmaking level, the extent to which indirect and transboundary effects would be incorporated into the decisionmaking process would depend

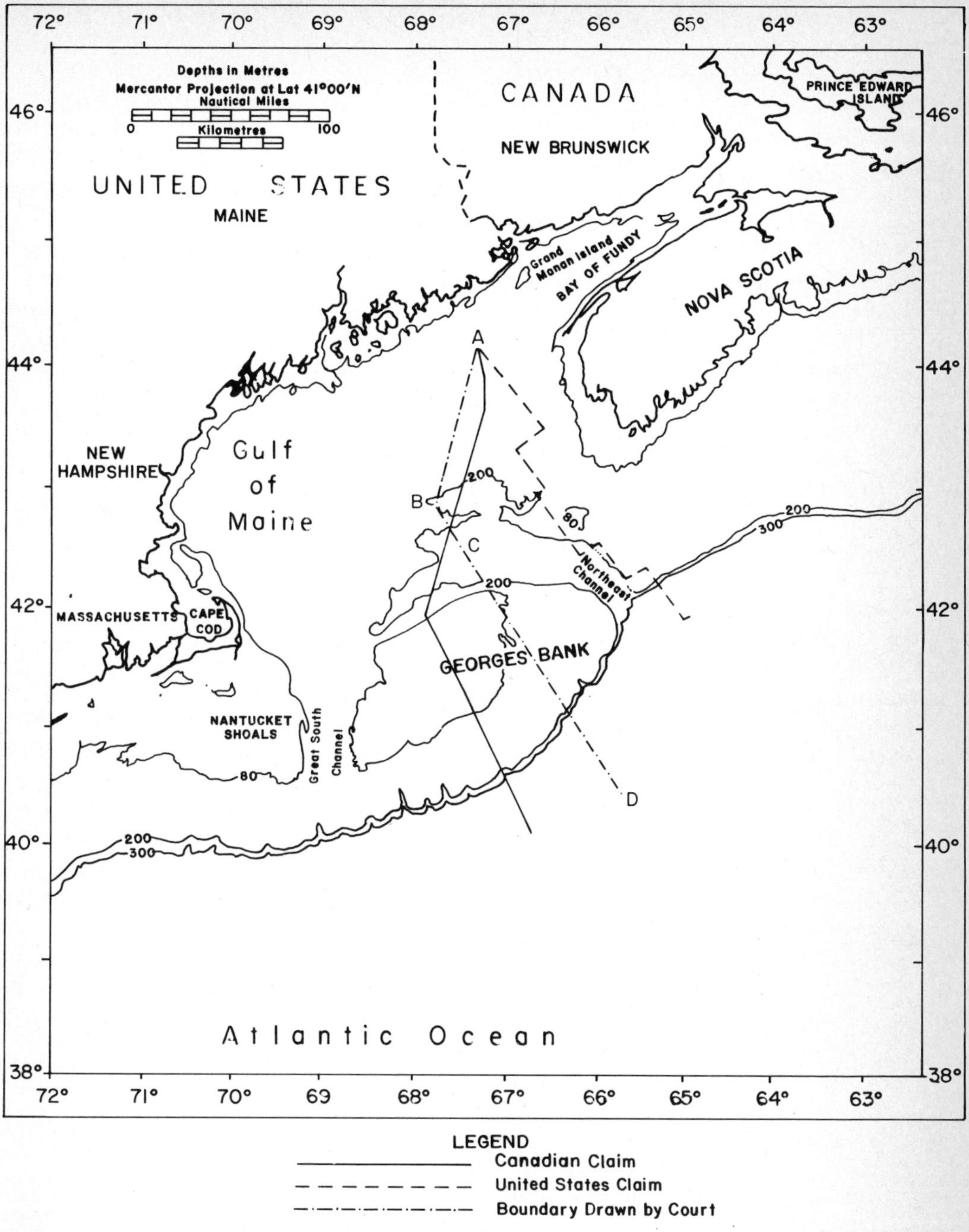

Figure 9. Gulf of Maine Maritime Boundary.

Adapted from Government of Canada, *News Release* 84/142 (October 12, 1984), p 3.

very much on official discretion.[1] A proponent of hydro-electric development in Nova Scotia would need to acquire various provincial permits including a water authorization pursuant to the Water Act. As part of the permitting process, the Minister of Environment could require an environmental assessment. Since the Nova Scotia environmental process is not set forth in legislation, regulations, or as a written policy, there is no guarantee that transboundary or indirect effects would be considered. The Environmental Assessment Division of the Department of Environment in its rather *ad hoc*, case-by-case approach could certainly require such analyses but much would likely depend on the extent of political opposition to a proposed project and the extent of extra-provincial concern. In New Brunswick the environmental assessment process is set out in a Government Policy adopted on October 8, 1975. It applies to all major projects that are government-sponsored or financed. However, the extent to which the assessment process would focus on indirect and transboundary impacts would depend on official discretion, for the Department of Municipal Affairs and Environment develops project-specific guidelines for the EIA or reviews guidelines prepared by the proponent.[2]

At the federal level, the Environmental Assessment and Review Process Guidelines Order does provide some guidance concerning transboundary and indirect effects. Section 4(1), added in 1984, urges special consideration of transboundary effects:

> An initiating department shall include in its consideration of a proposal ... the potential environmental effects of the proposal and the social effects directly related to those environmental effects, including any effects that are external to Canadian territory....[3]

This provision reflects a growing recognition that ecosystem boundaries frequently exceed political borders.[4] It likely stems from Canadian experience with Beaufort Sea oil and gas activities, controlling acid rain, and the proposed Fundy tidal power project.

Section 12(d) of the Guidelines may cover indirect effects by providing that where potentially adverse environmental effects of a proposal are unknown, the proposal must be subject to further study and subsequent rescreening or reassessment, or referred to the Minister of Environment for public review by an assessment panel. In case of panel review, the panel would have discretion as to the extent indirect effects would have to be considered.

U.S. National Environmental Policy Act

Environmental assessment in the United States is required by law under the National Environmental Policy Act (NEPA) of 1969. It applies to any "major Federal action ... significantly affecting the quality of the human environment."[5] NEPA is triggered by any activity involving a federal decision to approve, fund, or license activities carried out by others. Regulations developed by the President's Council on Environmental Quality (CEQ) specify the criteria for meeting that requirement as well as the scope, timing, preparations, and procedures for review of the environmental assessment.[6] In addition, each federal agency or department is required to implement regulations and procedures, subject to public review, to ensure conformance with NEPA.

The EIS, by examining the nature and extent of a proposed activity's environmental effects, is intended to ensure incorporation of NEPA's policies and goals in agency programs and actions. NEPA regulations specify that the EIS include a description of the affected environment and the environmental consequences of the activity on that environment.[7] Further, the unavoidable adverse effects, the means to mitigate those adverse effects, and the alternatives to the activity must be described.

Where insufficient information exists to assess impacts, CEQ regulations require the responsible federal agency to include a "worst case" analysis of the activity's potential environmental effects.[8] The analysis weighs the need for the action against the risk and severity of possible adverse impacts, if the action proceeds in the face of uncertainty, and estimates the probability that the worst case will occur. CEQ, however, proposed in 1985 to delete the worst case requirement, arguing that the analysis has in some cases been based only on speculation.[9] Under CEQ's proposed revisions, an agency faced with incomplete information would indicate where critical information is lacking and then base its analysis on what it identifies as "credible science'. The agency would then only evaluate impacts that were judged "reasonably foreseeable."[10] Impacts with low probabilities but catastrophic consequences would be included only if they were supported by credible science rather than "pure conjecture."[11] In its discussion of the proposed regulation, CEQ failed to indicate where the line should be drawn between credible science and scientific conjecture. Thus, interpretation of this regulation, as with other aspects of NEPA, will fall to the judiciary.[12] CEQ received close to two hundred public comments on the proposed revision, with support from major industry groups and criticism from environmental organizations. A principal concern is whether the revision would reduce agency incentives to conduct research which might reveal environmental problems.[13]

CEQ regulations also define indirect impacts.[14] They may occur later in time, far-removed from the activity, and affect air, water, ecosystems, historical or aesthetic features, the economy, social factors, or human health.[15] Relevant to weighing indirect impacts are requirements concerning controversial aspects of a project. Areas of dispute must be evaluated if the issues involve uncertain, unique, or unknown risks.

Expertise for addressing these issues comes primarily from the lead agency. Initial scoping sessions may also identify sources from other agencies. Meetings to fulfill coordination requirements under statutes like the Fish and Wildlife Coordination Act or from interagency agreements may also provide expertise. In addition, state and local participation as well as that of the applicant and interested and affected parties, initially solicited in the scoping stage, provides information. Comments on the draft EIS from the interested public and various advocacy organizations may also elicit information on indirect or transboundary effects.

The scope of NEPA has, in theory, been expanded by Executive Order 12,114, issued in 1979. This order requires federal agencies to evaluate the environmental effects of their proposed activities outside the geographical jurisdiction of the U.S. It arose from concern over the environmental effects of activities, particuarly those of the Departments of the Army and State, on foreign nations and the high seas.[16] The order provides among other things for various environmental studies, including environmental impact statements, bilateral or multilateral studies, and summaries of pertinent environmental issues.

Under the order, only those actions significantly affecting the global commons, where no domestic authority exists to make such an evaluation, trigger a full EIS. Here only effects on the *physical* environment, not social or economic impacts, need be considered. Thus, something less than an environmental impact statement is required for proposed activities potentially affecting a foreign nation. The U.S. is not required to assess the effects on another country of joint ventures between the U.S. and the other country. Failure to comply with this Executive Order would not result in litigation because the order precludes initiation of any action forcing compliance with its terms.

The principal shortcoming of NEPA probably is in the area of predicting the scope of impacts. Thus, indirect or transboundary effects remain the most elusive, difficult to document, and so difficult to mitigate. Overall evaluations of the process reflect this problem. Some see NEPA as providing no more than an *ad hoc* rationalization for agency action and seldom serving as an agency decision document.[17] Discontent with the process has resulted in extensive litigation. This judicial review has modified virtually every aspect of NEPA including definition of the statute, determination of what is sufficient federal particpation to trigger the process, and who has standing to challenge the adequacy of the procedures.[18] Completeness of the EIS or failure to meet any of the many procedural requirements under NEPA are common areas of litigation.[19]

Others argue that NEPA thwarts the “administrative passivity”[20] that previously characterized federal oversight of activities with environmental ramifications. Further, NEPA is viewed as playing an innovative role in public policy and environmental science. It has integrated science with public policy to an unprecedented degree, fostering an interdisciplinary approach within the scientific community to environmental problems.[21] Such an integrative approach can lay the foundation for considering ecosystem effects.

Conclusion

EARP and NEPA share the overall goal of assessing the environmental impacts of a proposed activity early in the decisionmaking process as well as considering the alternatives to the proposed activity. Several significant differences, though, exist between them. These differences strongly reflect the Canadian and U.S. styles of governance.[22]

The EARP relies heavily on negotiation. NEPA review is guided by specified procedures with the attendant potential for judicial review. The success of EARP seems to rest on the good judgment of the assessment panel, the proponent, and the Ministers of the Environment and the initiating department. Under NEPA, regulations depend primarily on procedural controls to ensure that science is sufficient and is incorporated in the decisionmaking process. These procedures form a network with other environmental protection statutes like the Clean Water Act and the Endangered Species Act (ESA), introducing additional procedures.

As for the actual assessment document, EARP procedures leave issues of content primarily up to the assessment panel. The document is prepared by the proponent, following the panel's guidelines. U.S. procedures provide detailed requirements for the document's content. They stress the importance of evaluating indirect effects and assessing the adequacy of scientific information. The EIS is usually prepared by the agency with principal or initial permitting authority.

Both processes now provide for consideration of effects outside territorial boundaries. EARP guidelines are fairly explicit, suggesting that where transboundary impacts may be present they be fully considered. U.S. requirements under Executive Order 12,114 actually amount to something less than the impact statement required for domestic activities. Evaluation is limited to assessing effects on the physical environment.

Utilization of Science in the Assessment of Indirect and Transboundary Impacts

NEPA provides greater specificity than does EARP for evaluating indirect effects. This specificity suggests that NEPA may be more sensitive than EARP in addressing indirect effects in transboundary regions. On the other hand, EARP guidelines urge special attention in areas where transboundary impacts may arise. Three projects, one in the U.S. and two in Canada, provide some basis for judging whether or not one system more successfully deals with indirect transboundary effects. These projects include the oil refinery and deepwater port proposed by the Pittston Company for Eastport, Maine in the U.S., exploratory oil and gas drilling in the South Davis Strait in northeastern Canada, and the Fundy tidal power project in Nova Scotia, Canada.

In addition to assessing the effectiveness of EARP and NEPA prescriptions in evaluating ecosystem information on transboundary indirect impacts, several more general categories will be considered here as a basis for judging the quality of actual assessments. These include sources for scientific and technical information, methods for integrating that information, bases for determining adequacy, and the characterization of uncertainty. Each of these will be considered in the context of the case studies.

The Pittston Oil Refinery Proposal[23]

The Pittston case illustrates several facets of the environmental assessment of indirect effects in a transboundary context. First, it demonstrates the degree to which early decisions concerning the extent of potential effects gain momentum and often narrow subsequent analyses. Agencies acting without the benefit of interagency consultation are particularly prone to this problem. The Pittston proposal also shows the difficulties of evaluating effects when the resources, both ecological and human, of another country are involved. It points to the importance of informal scientific exchanges among experts and the problems created when that exchange is impeded. Finally, it indicates the importance of a willingness to consider environmental uncertainty beyond what is required by law.

Exploration of the degree to which indirect, transboundary effects are considered and influence environmental assessment requires consideration of several factors. The first involves actual regulatory requirements that may specify the information to be collected in weighing project evaluation. Further, the adequacy of the referenced scientific and technical information to delineate impacts must be examined. Finally, why attention is given to some issues rather than to others needs assessment.

In 1973, the Pittston Oil Company proposed to locate a 250,000 barrel a day oil refinery and deepwater port for very large crude carriers (VLCCs) near the U.S.-Canadian border in Eastport, Maine (Figure 10). Pittston withdrew its proposal in 1983 only after the Maine Supreme Court ruled that the company did not have adequate title to the proposed refinery site.[24] Over the intervening ten year period, national priorities shifted. When Pittston was first proposed, the U.S. was actively promoting expansion of its refining capacity because of worldwide fuel shortages. By 1983, oil was again an abundant, relatively inexpensive energy source. Thus, the economic advantage of VLCCs had diminished, and so the Pittston proposal had lost its attractiveness.

Yet, changing availability of world energy supplies played little role in the debate over the refinery. Rather the intervening ten years were spent arguing over the potential effects of a catastrophic oil spill on the area around Eastport, Maine. The resulting adjudication involved the final environmental impact statement's (FEIS) characterization of the project's oil spill potential, particularly its effects on the region's endangered species.[25] The EIS, issued in draft form[26] by the U.S. Environmental Protection Agency's (EPA) Region I in Boston, Massachusetts, in October 1976, and in final form[27] in June 1978, was prepared as part of Pittston's application for a wastewater discharge permit under the requirements of the Clean Water Act.

Because the discharge permit was the first federal permit the Pittston Company applied for and a new source discharge under the Clean Water Act, the Council on Environmental Quality determined it to be a major federal activity requiring preparation of an EIS. Other permits subsequently involved the U.S. Army Corps of

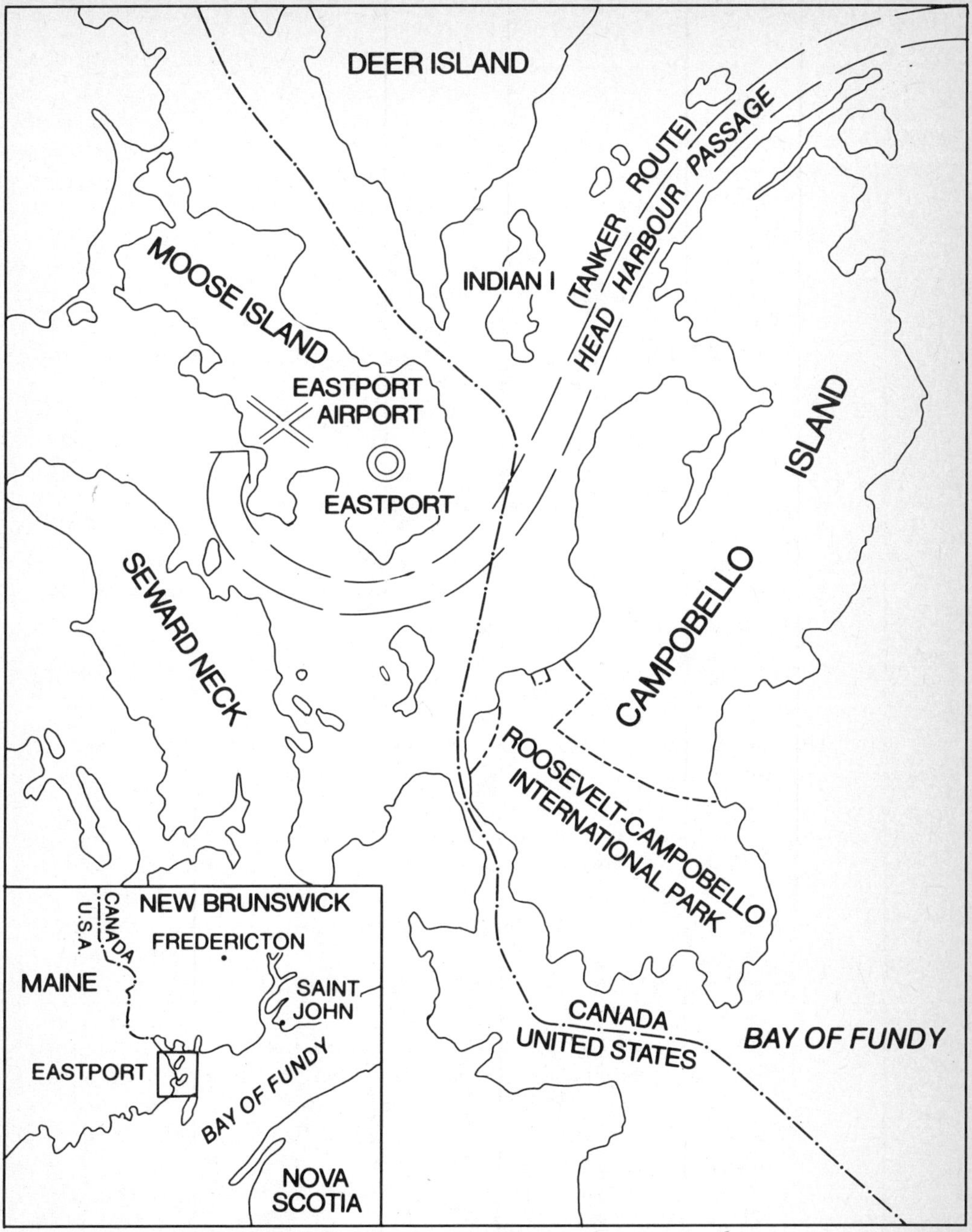

Figure 10. Proposed Shipping Route to Eastport Refinery Through Canadian Waters.

Adapted from map attached to Government of Canada, Head Harbour Passage Oil Carriage Limitation Regulations - Background Notes.

Engineers for dredging and pier construction and the Federal Aviation Agency for rights to the refinery site, formerly an airport. In conjunction with the U.S. EPA, the agencies designated U.S. EPA as the lead agency responsible for EIS preparation.

U.S. EPA's EIS concluded that a major oil spill from the project was unlikely and that it therefore posed little danger to the region's resources.[28] The Agency had to reverse this decision, however, after consultation procedures under the Endangered Species Act (ESA) determined that the Pittston operation would jeopardize the area's endangered species.[29] In 1981, the Pittston Company, however, successfully appealed U.S. EPA's permit denial in an adjudicatory hearing.

This hearing on the Pittston discharge permit determined among other things that: the FEIS complied with NEPA, the risk of an oil spill was minute, and operation of the Pittston refinery was unlikely to jeopardize endangered species.[30] Therefore, U.S. EPA was required to issue Pittston a permit. In turn, this permit issuance was challenged in the First Circuit Court of Appeals by the Conservation Law Foundation, the Roosevelt Campobello International Park Commission, and several other groups.[31] In that litigation, the judge ruled that U.S. EPA's determination of no jeopardy to endangered whale species had to be reconsidered using real-time simulations of tankers transitting Head Harbor Passage, hydrographic surveys of those waters, and relevant ecological studies not previously considered or completed at the time the EIS was issued. Only then would U.S. EPA have fulfilled its obligation under NEPA and the ESA to use the best scientific data available.

The Canadian government expressed immediate opposition to the Pittston proposal. Underlying its opposition was Canada's belief that transporting a hazardous substance, oil, through its internal waters was illegal without federal government concurrence. Further, Canada contended throughout that diplomatic priority disallowed Canadian officials from engaging in any U.S. proceedings on the issue.[32] It held these positions over the ten year period of the project's consideration.[33] In addition, both Canada's Ministry of External Affairs and the U.S. State Department maintained that discussion of the project's potential transboundary effects should be their concern, not that of federal agencies responsible for domestic regulation.[34]

Moving consultation to the Cabinet level internationalized project deliberations and precluded Canadian participation at several critical junctures. For example, in 1973 when Canadian marine pilots experienced in navigating the area were invited to testify in the Quoddy region before the Maine Board of Environmental Protection's (BEP) permit hearings, their participation was refused by the Canadian government.[35] Thus their testimony and that of scientists from Canada's Department of Fisheries and Oceans laboratory on the Passamaquoddy Bay was precluded at EIS hearings.

Because it lacked the Canadian expert testimony, Maine's BEP conditioned its permit issuance on Canadian agreement about transit and pilotage procedures and on navigational aids in Canadian waters.[36] U.S. EPA, however, proceeded without direct Canadian advice. Perhaps it assumed that the U.S. State Department would have access to the appropriate information and, when it became available, would channel it to U.S. EPA.

The persistence of some Canadian and U.S. officials and scientists did on occasion override federal government strictures. An August, 1976, meeting between Canadian and U.S. representatives held to exchange information on the Pittston project broke down because of Canada's unwillingness to join in U.S. hearings on this issue.[37] By coincidence though, Canadian and U.S. representatives in each delegation were also appointees to a fisheries committee of the bilateral International Joint Commission. Under the auspices of this organization, the meeting resumed. When the Department of Commerce's National Oceanic and Atmospheric Administration (NOAA) attempted in 1979 to obtain data on the region's marine resources, normally available on a scientist-to-scientist basis, it was only able to acquire that information after Canadian authorities were convinced to publish it in a Fisheries and Marine Service Technical Report.[38]

Further, officials from Transport Canada, Fisheries and Oceans, Environment Canada, and New Brunswick's Department of Environment did comment on U.S. EPA's draft EIS. They raised serious questions about the effectiveness of Pittston's proposed navigation systems, the EIS's description of the Head Harbor Channel's morphology, and the characterization of the project's oil spill potential.[39] Moreover, scientists from the Department of Fisheries and Oceans prepared several reports assessing the risks of oil spills on the Passamaquoddy region.[40] Little of the information in these reports found its way into the EIS draft although the reports proved useful in subsequent litigation.

Ultimately, Canada made formal its position that transport of oil through Head Harbor Passage violated its laws. In 1982, it adopted the Oil Carriage Limitation regulation under the Canada Shipping Act.[41] This regulation prohibited oil tankers carrying more than 5,000 cubic meters of oil from transitting Head Harbor Passage. (The amount was estimated to be equivalent to the fuel necessary to meet the Eastport area's hydrocarbon needs.) Under the Canada Shipping Act, the government possessed authority to restrict shipments of hazardous cargoes through its internal waters. The Canadian government determined that petroleum shipped through Head Harbor Passage was a hazardous cargo. Its declaration seemed to offer little room for negotiation concerning the project; yet, the Pittston Company, possibly assuming that U.S. federal intervention could affect this determination, proceeded with its applications.

Ironically, Canada's posture may have complemented Pittston Company's strategy for dealing with Canadian opposition. The company perceived that any Canadian advice on the project would be negative and would hurt public support in the U.S. for the project. It is rumored that Pittston did not solicit that advice from Canada in order to allow project support to build in the U.S.[42] Pittston apparently reasoned that subsequently U.S. domestic support would overwhelm any Canadian objections.

Any discussion of the proposed Pittston project must consider several caveats. Concern in the 1960s about human encroachment on the environment gave rise to extensive environmental legislation including the National Environmental Policy Act, the Clean Air Act, amendments to what is now the Clean Water Act (CWA), and the Endangered Species Act (ESA).[43] In the 1970s, these laws and their implementing regulations underwent considerable refinement.

The Pittston refinery was among the early projects to be subject to the new environmmental legislation, and the project EIS was among the first prepared by U.S. EPA's Region I under the provisions of the CWA. NEPA requirements as well as those of other state and federal statutes created what has been described as a regulatory maze.[44] As the legislation was amended, the maze became more complicated. The Pittston Company, thus, had to respond to changing requirements.

Application of the Endangered Species Act's consultation provisions, triggered by NEPA review, illustrates the difficulties encountered with this rapidly evolving legislation. Several key decisions on Pittston's discharge permit and, ultimately, the legality of the entire project, turned on its potential to harm the area's endangered species, among them bald eagles and right whales. Enacted in 1973, the ESA under its Section 7, mandated consultation on an activity's potential to threaten endangered species. Participating agencies were the Department of Commerce's National Marine Fisheries Service (NMFS) and the Department of Interior's Fish and Wildlife Service (FWS) and the lead agency for the federal activity - in the Pittston case, the U.S. EPA. No formal procedures for consultation were specified until final regulations were issued by NMFS early in 1978, five months before the final Pittston EIS was released.[45] Further, definition of key words such as "jeopardy" and "critical (habitat)," specified in Section 7 to be used in defining potential harm, remained undefined until the final regulations were issued.

U.S. EPA initially interpreted "consultation" under Section 7 as advisory.[46] It viewed its obligations as fulfilled by preparing the EIS with its conclusion that the project presented no threat of adverse impact on the endangered species, and distributing the EIS to NMFS and FWS.

In 1977, NMFS published ESA guidelines, six months before U.S. EPA released the draft EIS.[47] These guidelines set forth formal consultation procedures, making ESA consultation procedurally bound with NEPA. They were essentially similar to the Section 7 regulations adopted in 1978 and were upheld upon judicial review.[48] Further, soon after issuance of the final EIS Congress amended the ESA, providing a detailed, formal consultation procedure essentially similar to the 1978 regulations. Following the implemented consultation procedures, U.S. EPA, as noted above, concluded that the project threatened the region's endangered species and denied Pittston's permit, reversing the conclusion of its FEIS.

Thus, during the ten year span of the Pittston project, applicable provisions like those under the ESA changed, in some cases substantially.[49] Nonetheless, the themes raised by Pittston - narrow interpretations which limit subsequent considerations, reliance on limited data without specification of its limitations, and failure of interagency coordination - still arise in the U.S. environmental assessment process, suggesting that the issues Pittston raises are generic rather than idiosyncratic.

The debate about the Pittston project revolved around three related issues - navigational safety, fate of spilled oil, and impacts of spilled oil on endangered species. The Passamaquoddy's waters are among the foggiest in the world.[50] Tanker passage to the proposed refinery site involved transitting the Canadian waters of Passamaquoddy Bay through Head Harbor Passage which at one point narrows to under one mile. The region's waters support important commercial populations of groundfish, herring, and lobster.[51] Finally, the waters outside Passamaquoddy Bay are the summer feeding and breeding ground for right whales and support year round populations of bald eagles, both endangered species.[52]

The difficulties with the proposed refinery site were highlighted in a letter from a member of EPA Region I's Environmental Impact Office to the Regional Director following the former's visit to Eastport with representatives of the State Department and the U.S. Army Corps of Engineers.[53] A trip through Head Harbor Passage was terminated after the rapid onset of fog; however, the trip was of sufficient duration to indicate that during foggy periods tankers would be operating with zero visibility and that the Head Harbor channel provided no appropriate anchorage. The U.S. EPA representative, along with those from other federal agencies, agreed that other sites along the Maine coast would be more appropriate for a deepwater oil tanker port.

Oil Spills. Region I's attempts to assess the potential effects of an oil spill in the Passamaquoddy region illustrate the way in which assumptions made early in the decisionmaking process tend to narrow subsequent considerations. Initially, the Agency incorporated an analysis provided in Pittston's environmental assessment report (EAR).[54] The U.S. EPA, the then-Federal Regional Council (a regional and federal group to encourage energy development in New England), and the Pittston Company determined the need for an EAR because of the complexity of the information involved.

The approach in the EAR focussed on oil transfer operations as being the major sources of oil spills and so assumed that the major danger from oil spills occurred at the docking and transfer facilities rather than in transit.[55] It argued that statistics from a similar facility located at the Welsh port of Milford Haven provided a basis for predicting that limited spillage would be small at the proposed Pittston facility. The EAR data indicated that 90 percent of the spills at Milford Haven occurred when the tankers were berthed, and of these, 90 percent involved less than 4 barrels (168 gallons). Only 3.7 percent occurred during tanker passage or at moorings. Over a nine year period of operation, the average percentage of oil spilled to oil handled was exceptionally small (0.00068 percent).

The draft EIS indicated that based on experience at Milford Haven, chronic spillage from routine transfer would be between 20 and 86 barrels per year, and an additional 1-2 barrels per day would be discharged in the wastewater.[56] No analysis of the cumulative effect of these wastewater discharges was made. Finally, using the EAR data and noting that the tidal range, currents, and channel characteristics at Milford Haven were "all similar to Eastport in many ways,"[57] the draft EIS concluded that the Pittston facility could be expected to be safe and clean.

Responding to criticism of its oil spill analysis in the draft EIS, the EPA in the final EIS constructed scenarios to estimate the probability and extent of damage for spills from non-berthed tankers in locations around Cobscook Bay, Passamaquoddy Bay, Head Harbor Passage, and the eastern shore of Campobello Island. The scenarios ranged from a "worst-case" 80,000 gallon crude oil spill (100 percent of tanker capacity) in Head Harbor Passage to 3,000,000 gallons of –2 fuel oil (14 percent of capacity) spilled at the mouth of Cobscook Bay.[58] Variables included spill location, volume, oil type, intertidal zone size, tidal range, and current velocity.

This approach allowed EPA to make quantitative estimates of the geographical and ecological extent of damage from a spill in contrast to the qualitative approach in the Milford Haven example. The draft EIS acknowledged that for the spills occurring in the Head Harbor Passage (three of the five scenarios) 50 percent of the oil would be transported beyond the Quoddy region to the eastern shore of Campobello Island.[59] Yet, EPA failed to pursue this contention, apparently following the reasoning in the draft EIS that the greatest potential for both accidents and serious damage lay with spills closer to the refinery site.

Part of the difficulty arose from a failure to account for the role of seasonally variable winds and the differential effects surface and subsurface currents play in transporting oil. Prevailing winds in this region shift from south/southwest in the summer to north/northwest during October to March.[60] Thus, oil from a spill at the mouth of the Passamaquoddy could be blown by winter winds along eastern Campobello Island down the Maine coast. A summer spill in the same location, however, would drift in surface currents to the mainland coast of New Brunswick and the surrounding islands. Subsurface currents would still carry oil southward.

This difference in behavior between winter and summer spills was documented in oil spill modeling for the Georges Bank and Gulf of Maine, begun at the University of Rhode Island in 1974 and applied to the Eastport region in 1979.[61] The modeling combined long term data on winds, currents, and temperatures in its three-dimensional mass transport model. Simulations of multiple trajectories were conducted for a winter and a summer spill of 13 million gallons, located 2.5 miles northeast of Campobello Island. Thus, the 1,000 simulations run for each spill showed the changing transport of the oil with daily shifts in wind and currents over a fifty day period. Single event spills were also modeled using historical data on daily wind shifts and currents.

Modeling indicated that a winter spill at the entrance to Passamaquoddy Bay would be transported by the prevailing north/northwest winds southward into the Quoddy region, offshore of Grand Manan Island, along the coast of Maine, and eventually toward western Nova Scotia. In contrast, summer winds would drive oil into the Passamaquoddy and up along the mainland coast of New Brunswick. Subsurface flow, largely unaffected by wind, would move oil southwest.

The EPA concluded in its final EIS that an oil spill was extremely unlikely and that if it occurred, would result in only a small localized amount of damage. This conclusion appears to have driven EPA's reliance on its scenario approach. However, discussion concerning environmental conditions in the final EIS focused on conditions within the Passamaquoddy Bay, reinforcing the idea apparent in the draft EIS that the spills would only occur close to the refinery site.[62] Neither the difficulties in characterizing hydrographic conditions in the highly complex Passamaquoddy-Fundy region nor the implications of the limitations, inadequacy, and shortcomings of data on meteorology, tidal currents, and residual (net non-tidal water movement) current conditions were considered in the EIS.

Reference to modeling and other studies would have shown daily and seasonal changes in wind and current. These changes could carry oil out of the Passamaquoddy Bay. Data easily obtainable (for example, on winds from the Portland, Maine, National Weather Service Office) or in the scientific literature (regional wind patterns,[63] currents,[64] and local winds and currents)[65] would have suggested to EPA, even taking into account their apparent view of a localized spill, that they should expand the geographic and seasonal scope of its analysis. Why they did not seek out this additional information and were unaware of the modeling research being done locally in Rhode Island is unclear.

Navigation. U.S. EPA's narrow view of an oil spill's potential extent had significant implications for the two other critical areas, navigation and endangered species. EPA's contention that the probability of an oil spill would be small came in part from the Pittston Company's description of the navigation aids it intended to install to assure safe transit. While Head Harbor Passage and Friar Roads provide a deep natural access channel leading to Eastport, the large tidal range, variable currents, and extreme weather conditions commonly resulting in fog, make the twisting, narrow channel difficult to navigate.[66] These unique conditions require special attention to navigational safety.

In considering Pittston's proposal, numerous questions were raised concerning the safe navigation of VLCCs and smaller product vessels through these waters. The Canadian Department of Transport first raised its objections in 1973.[67] These objections were reinforced in several other reports that considered weather conditions, uncertainty over channel widths and depths, and the absence of a safe anchorage once transit began.[68] These documents directly challenged Pittston's description of the waterways and of the proposed navigational aid's effectiveness.

Faced with Pittston's and Canada's contradictory judgments and denied access to Canadian pilots with knowledge of the the area, the U.S. EPA deferred to the U.S. Coast Guard's expertise. The Coast Guard indicated that the passage could be safely navigated if four conditions were met.[69] These were new hydrographic surveys of the channel; a monitored navigation system; a mechanism to steer tankers in case of steering or propulsion loss; and strict operating procedures. The final EIS, citing the Coast Guard's appraisal, concluded that the necessary navigational safety program could be implemented.[70]

In the subsequent litigation, the Coast Guard clarified its position, arguing that EPA had misinterpreted its comments.[71] It maintained that it had actually meant only that it was willing to devise aids to assure minimal navigation risks. In the Coast Guard's view, the risk analysis implied by EPA's interpretation of the Coast Guard's position would only result after real-time simulation studies, ballasted VLCC test runs, and up-to-date hydrographic surveys had been conducted. Even with these, the U.S. Coast Guard suggested that fog conditions could preclude safe transit.

Endangered Species. U.S. EPA's narrow interpretation of an oil spill's potential in the Eastport area also constrained its analysis of possible effects on the region's endangered species, particularly marine mammals. Data, often only anecdotal, available on these animals' distributions, abundances, and migratory behavior was too limited to overcome the flaws in the oil spill analysis. Finally, consultation procedures among the affected federal agencies intended to identify shortcomings in the environmental analysis failed. The failure arose because of conflicting interpretations of ESA's procedural requirements by the affected agencies.

The EIS lists seven of the area's species (five whale species and two bird) as endangered under the ESA.[72] The document concentrated on the effects of oil on right whales, considered most susceptible to oil.[73] Further, it recognized that fuel oil with its soluble and therefore, largely invisible, fractions posed a greater threat to marine organisms than did crude oil. EPA contended that because of their mobility most whales would be able to avoid a crude oil spill and that the low probability of an oil spill limited any exposure to refined hydrocarbons.[74]

Comments by a whale expert indicated that right whales fed on and just below the surface and so would be particularly susceptible to an oil spill.[75] The final EIS did not reflect this information.[76] This omission is not surprising because the agency's evaluation of marine mammal effects was structured by the scenario analysis. The scenario approach focused on oil spill effects only within the Passamaquoddy while distribution studies showed that right whales summered just outside the Passamaquoddy.

Consultation provisions under the ESA were intended to correct the inadequacies like those in EPA's analysis. Activation of the Section 7 provisions, however, was impeded at several stages. The National Marine Fisheries Service, in commenting on the draft EIS, did not raise the consultation issue. It did, however, note that the document lacked sufficient information to assess impacts "on living marine resources."[77] Insufficiency under ESA's Section 7 usually indicated that consultation should be initiated to evaluate the proposed permit.

Endangered species issues were, however, discussed by the critical resource protection agencies. For example, in May 1976, five months before the draft EIS was issued, representatives from NMFS, FWS, EPA, and the Federal Regional Council's New England Refinery Task Force met to discuss endangered species issues.[78] At this meeting, EPA was urged to contact the NMFS Northeast regional office for additional information on marine mammals.

A request from Region I in June 1978 after the final EIS was issued, specifically asking for NMFS comments on endangered species, finally galvanized NMFS to initiate consultation.[79] NMFS' threshold Section 7 determination, conducted during the fall of 1978, concluded that information was insufficient to determine if the project would jeopardize the area's endangered whale species. At the same time, FWS judged that the project would jeopardize the local population of bald eagles. A finding of jeopardy under the ESA precludes federal permit issuance. Therefore EPA, despite the conclusion of its EIS, was forced to disapprove Pittston's permit application.

The EPA's failure to apply full consideration seemed to have resulted from its interpretation of consultation as advisory rather than requiring a more active exchange of information. Apparently, EPA headquarters in Washington did not apprise the region of legal developments in the interpretation of the ESA. Less clear in light of ESA guidelines are the reasons for NMFS' failure to initiate consultation until after the EIS was issued. NMFS' comments on the draft EIS did not raise any endangered species questions. The document may not have been routed through the Endangered Species Office or it may not have been reviewed by someone with endangered species expertise. Both only become requisites after implementation of the final ESA regulations.

Review Under the Canadian Environmental Assessment Review Process (EARP)

Under EARP, evaluation of scientific adequacy falls largely into the hands of the assessment panel. This group, utilizing its knowledge of the available information's quality, both sets guidelines for the EIS and weighs the degree to which they have been met in that document. Choosing specialists for the panel from the relevant areas should insure that the panel addresses the diversity of issues involved. Panel recommendations become critical in the identification of far-field, indirect impacts. Thus, any analysis of the utilization of science in Canadian environmental assessment must focus on the interaction of the panel with that process.

Two sources provide information on how the EARP process and the assessment panels work in evaluating far-field, indirect impacts. First, a recent review based on a larger study of how ecological information is used in the Canadian environmental assessment process examines the role judgments about scientific adequacy play in those assessments.[80] The larger study supplies an overview of how effectively ecological information, some involving far-field or indirect effects, is incorporated into EARP determinations. In addition, examination of one project in particular, the South Davis Strait Petroleum Exploration Project, allows a more detailed view of how information on far-field, indirect effects is employed in the assessment process.[81]

Ecological Information in Environmental Assessment

The goals of public policy and science differ.[82] This difference often creates tension when science is applied to the decision process involving technology, particularly where uncertainty is present. While the former leads to one unequivocal resolution, the latter through an iterative process of testing seeks to acquire new knowledge by discrediting the old. Thus, decisionmakers and scientists are often unhappy with results of the application of science and scientific expertise in public policy. The greater the uncertainty concerning the scientific data - a feature which characterizes far-field, indirect effects - the greater the problems in applying science.

Beanlands and Duinker address this problem in their study.[83] They reviewed twenty-one EISs and panel reports prepared for eighteen representative projects. Of those projects, twelve were judged by the Department of Environment acceptable to proceed, two were still under consideration at the time of the study, and four had been rejected. Of those rejected, two (Lancaster Sound offshore drilling and Roberts Bank port expansion) were on ecological grounds and the other two (both Eldorado uranium refinery proposals) for socio-economic reasons.

For those assessments studied, ecological information only stood as a basis for project rejection when great, demonstrable ecological, social, economic or cultural loss would occur or when a high risk to those resources could be shown. Seldom were data complete or conclusive enough to establish the potential or the high probability of irrevocable loss. These deficiencies in the ecological data may explain why it is seldom the exclusive basis for rejecting a project.

The panel reports reviewed commonly noted that data were insufficient to assess the project's impacts. In their evaluation of the EISs, these panels also often observed that important data were still missing. Nonetheless, as the proportion of projects accepted-to-requested indicates, panels seldom viewed the shortcomings of the data as sufficient basis for project rejection.

Ecological data, while only rarely serving as the basis for project rejection, was nonetheless critical in the panel guidelines and so played an important role in shaping the actual environmental assessment. Most of the EISs examined by Beanlands and Duinker, however, were found to lack sound scientific information as a basis for assessing potential project effects. The information was either excessive, uninterpreted, or presented extensive verbal descriptions of the environment, lacking a quantitative basis.

Part of the fault derived from the generality of panel recommendations, which attempted to cover all possible impacts. General panel guidelines yielded general assessments. Considerations phrased as hypotheses to be tested or as specific questions, in contrast, provided the proponent something specific with which to work. They also supplied the panel with a basis for judging the degree to which the EIS fulfilled its criteria.

Assessments judged to consider adequately ecological factors defined boundaries early in the project. These included administrative boundaries (the time and space limitations that political, social, and economic forces play in the assessment); project boundaries (the time and space scales over which the project operates); ecological boundaries (the time and space scales of the natural system); and technical boundaries (limits placed on abilities to predict or to measure ecological change). The confluence of these boundaries defines the universe for assessment procedures, and promises to include indirect effects as well as those more obviously observed and measured.

South Davis Strait Offshore Petroleum Project

The proposal to conduct exploratory drilling in South Davis Strait was initiated in 1976 by three oil companies, Imperial Oil Limited, Aquitane Company of Canada Limited, and Canada Cities Service Limited. South Davis Strait is located in northeastern Canada and bounded by northern Labrador, Baffin Island, and Greenland. The companies proposed a joint assessment for the activity to the Department of Indian Affairs and Northern Development (DIAND). DIAND requested that the assessment be conducted as a regional study rather than a site-specific, well-by-well analysis of potential impacts.[84]

Guidelines for the EIS were first issued to the proponents by DIAND in 1976. Formal guidelines were released in 1978, the same year as the EIS. The proponents and their consultants did, however, have access to a draft copy of those guidelines in the later stages in the EIS preparation.[85]

Insufficiencies in the EIS resulted in a supplement being prepared in 1979.[86] The initial EIS's shortcomings arose in part because EARP was applied so late in the planning process and because the guidelines were overly general.[87] For example, the guidelines urged identification of all possible impacts rather than focusing on critical changes which could be documented. Their broad approach made it difficult to determine whether the EIS was satisfactory. Despite agency dissatisfaction with the proponents' assessment, final project approval was granted in 1979.

Project Assessment Objectives

The panel objectives were "to determine those areas where, from an environmental point of view: (i) drilling can proceed and under what conditions, (ii) drilling cannot proceed, and (iii) insufficient data exists (sic) on which to base a decision."[88] In contrast, the proponents saw their role as "defin(ing) the possible impact that drilling activities may have upon the offshore region and the adjacent Baffin Island area."[89]

In their analysis of the South Davis Strait assessment, Beanlands and Duinker suggest that the proponents were unlikely to follow the Federal Environmental Assessment Review Office (FEARO) guidelines for two reasons.[90] First, each had specific sites for which it sought approval and so was unlikely to reject its site in favor of another. Second, proponents were seeking approval for their projects and were unlikely to highlight areas which could cause delay or project disapproval. Whether or not this was the case, the objectives provided little specific guidance in developing the assessment.

Incomplete Information and the EIS

The proponents, in preparing the EIS, basically followed their view of what should constitute the assessment. Little information existed at the time on the effects of oil and gas activities in northern marine environments. To correct this deficiency, the DIAND, in conjunction with the Departments of Environment (DOE) and Energy, Mines and Resources (EMR), established the Eastern Arctic Marine Studies Program (EAMES) in 1977.[91] EAMES coordinated environmental studies in Davis Strait as well as in Baffin Bay and Lancaster Sound.

The early EAMES studies, supported in part by the proponents, were conducted in the South Davis Strait area. They concentrated on gathering baseline information, mainly on the distribution and abundance of organisms in the area. The results of such studies are difficult to use in environmental assessment unless they contain measures of natural variability. Further, the logistical difficulties in sampling in northern marine waters often preclude collection of this type of information. Thus, the results often only provide the descriptive data commonly found in EISs.

Field data from two years of sampling was available for use in preparation of the EIS. Additional information on the 1978 field season was incorporated into the supplemental EIS. Information, however, resulting from the International Biological Program on the region's sensitive areas had been collected but was not referenced in the EIS. Nor was hydrocarbon toxicity data on cold water marine organisms which had been compiled for other Arctic regions.[92]

Impact Identification

Interviews with proponents and consultants indicate that in evaluating impacts for consideration in the EIS an informal matrix of potential impacts and effects was constructed.[93] This matrix listed project activities and likely events along one axis and the environmental components along the other. The expected interaction between any two elements could thus be evaluated.

The EIS distinguished between major, moderate and minor impacts. A major impact, for example, was judged to affect an entire species, lowering its abundance over the longterm and removing it as a commercial resource. A moderate impact would affect a species' abundance or distribution over one or two generations but only to the extent that it eventually recovers. This approach along with the informal interaction matrix allowed priorities to be determined in order to evaluate the significance of potential impacts.

The proponents judged that the major environmental threat from the project came from the possibility of an oil or gas blowout.[94] Other considerations such as the effects of normal operations or rig servicing were determined largely to be unimportant. Thus, baseline data for the region and information from the literature on hydrocarbon toxicity were evaluated based on the results of oil spill trajectories developed using computer simulations.

Modeling results suggested that an oil spill would move south and westward from the area of drilling.[95] This information resulted in expansion of the assessment to include the Hudson Strait, Ungava Bay, and the Labrador coast. Unfortunately, because the modeling effort came after the design of the field program, the expanded area vulnerable to a spill was not included in the field studies.

Some studies conducted as part of the EAMES program and utilized in the EIS examined community level effects for plankton in particular.[96] These provided documentation that there was a high level of spatial variability in the spring plankton bloom, and this bloom contributed significantly to biotic productivity during the rest of the year. This research also contributed to understanding the region's trophic structure because of the central role the plankton played in supplying nutrients. Based on these studies it was possible to assess at what time of the year greatest damage from an oil spill might occur to the plankton populations.

Generally, the scope of research used in the EIS was limited for two reasons. First, so little basic information was available on the Davis Strait ecosystem that this foundation knowledge had to be collected before more sophisticated studies could be pursued. Sufficient time, however, between project initiation and approval was not available to complete both types of research. Second, because the principal impact of the activity was judged to be from an oil spill, analysis concentrated on that facet and studies were limited to what were considered to be a few important species.

In summary, predictions were based on two seasons of field studies, literature on the effects of oil on species, and the oil spill trajectory analysis. Based on their magnitude, extent, and duration, impacts were classified as major, moderate, or minor. This analysis indicated that while the major threat stemmed from an oil spill, the probability of that event was extremely small and posed little threat to the region's otherwise vulnerable resources. The assumption remained that there was only a low probability of an oil spill.

One example from the EIS cited by Beanlands and Duinker illustrates how the proponent's line of reasoning constrained its evaluation of vulnerable populations.[97] During late winter, polar bears occupy ice floes around the Davis Strait. Based on limited studies, the EIS concluded that an oil spill would result in only moderate damage to the local polar bear population. The proponents adopted the strategy that where data were insufficient, as was the case for polar bears, a worst case analysis would be applied. This technique meant that the impact would be upgraded. Thus, the Davis Strait polar bears were considered to be subject to a major impact. The EIS, however, arguing that the probability of an oil spill was small, maintained that despite the bears' vulnerability the threat to them was minimal.

Agency Response

The EIS met with considerable criticism from the two principal resource protection agencies in Canada. The Department of Fisheries and Oceans, in its review of the trajectory modeling effort, indicated that improper data on winds, tides, and currents were used.[98] The proponent argued that the worst case analysis used compensated for any shortcomings in the data.

In its review, the Department of the Environment identified deficiencies in the EIS's data on marine mammals, weather prediction, hydrocarbon levels in the water column, sediments, and animal tissues, and in the definition of sensitive areas.[99] EAMES data on the distribution of seals, fishes, polar bears, and birds were unavailable and so not included in the EIS.[100] Further, the Department of the Environment considered that the chance of an oil spill was greater in the South Davis Strait than in any other area of the Canadian offshore.[101] They also found that oil spill projections were based on incomplete or inaccurate meteorological, biological, and physical data. Finally, the Department of the Environment noted the absence of available data on sensitive areas collected under the International Biological Program.[102] The assessment panel, however, did not share these agencies' dissatisfactions and, in particular, found the modeling effort acceptable.

The Department of the Environment recommended that offshore activities be prohibited until a more comprehensive assessment was supplied.[103] The main resource exploitation agency, the Department of Energy, Mines and Resources, however, concluded that despite the EIS's shortcomings the project did not pose any unacceptable risks to the environment.[104] Finally, DIAND urged project approval, arguing that the proponents had a close consultative relationship with the government.[105] This relationship would provide a high level of assurance that problem areas could be addressed in future EAMES studies.

In their recommendations concerning the EIS, the assessment panel qualified their approval on several conditions.[106] These included instituting a real-time monitoring program; a detailed oil spill mitigation plan prior to the initiation of drilling; a government oil spill contingency plan delineating agency responsibilities; and a study examining the increase in liability for any damage and clean-up costs not covered under existing compensation plans.

Conclusion

Despite DIAND's initial injunction for a regional assessment, a narrow environmental review resulted. No consideration was given to the possibility of extra-national, indirect impacts on Greenland's resources. This limitation probably resulted because modeling suggested that an oil slick would move south and westward. Yet, both the resource agencies in their evaluations faulted the fundamental information that went into that effort. The mutually reinforcing assumptions that the principal threat came from the very small probability of an oil spill and in that unlikely instance an oil slick would behave as the modeling predicted imposed a narrow focus on the assessment. The additive effects of multiple drilling in the region from, for example, the nearby British Petroleum operations were not examined nor were those from adjacent drilling activities then proposed for North Davis Strait and Lancaster Sound.[107] Curiously, neither DIAND nor the assessment panel highlighted these shortcomings.

Both the EIS's and the panel's recommendations reflected a short term, near-field view of the activity's impact on the environment. This view also guided the design of the EAMES-sponsored studies. One critique of the Davis Strait review argues that uncertainty over requirements for EARP itself made it unclear whether the assessment was intended to identify areas of deficient knowledge or to justify the environmental impacts of the project.[108] Faced with the generality of the panel guidelines, the proponent proceeded with an assessment that met its objectives. The panel did recommend that project approval be delayed until the proponents presented a better depiction of possible impacts and available mitigation measures, though in light of the paucity of data on the area's resources only a major longterm study would have provided that information. The panel's more specific recommendations, involving establishment of monitoring and oil spill contingency plans, suggest that evaluation of the activity's impact on the region was seondary to moving ahead with the project.[109]

The EARP for Fundy Tidal Power Projects

The early stages of the assessment for the proposed Fundy tidal power project and the actual process applied to the Fundy prototype, the Annapolis Tidal Power Project, provide another source of information on how EARP deals with indirect effects and far-field effects. Further, Annapolis assessment was conducted by Nova Scotia's Department of the Environment (NS DOE). Thus, the Annapolis example suggests how assessment functions on a provincial level.

As described in chapter 1, a report issued in 1977 by the Tidal Power Review Board concluded that tidal dams using the upper Bay of Fundy's exceptionally high tides would be economically feasible.[110] The Fundy Tidal Power Environmental Assessment Panel, consisting of representatives of the federal government and that of New Brunswick and Nova Scotia, was then formed to draw up guidelines for an EIS. The resulting guidelines required consideration of the dam's effects on the Gulf of Maine although they primarily focused on the need to examine impacts in the upper Bay region.[111]

These guidelines provide a comprehensive basis for the preparation of an EIS. Since their issuance, however, considerably more information has been collected and synthesized on the upper Fundy ecosystem.[112] Also, computer simulation of the Bay of Fundy and Gulf of Maine tidal regime suggests potential effects unknown at the time the guidelines were written.[113] Thus, if the Tidal Power Corporation decides to move ahead with the project, new guidelines will doubtless be drawn up.

The Annapolis Tidal Power Project provides an opportunity to assess how questions of adequacy of scientific data were addressed by decisionmakers on the provincial level. Following guidelines developed by the NS DOE, an EIS was prepared by a consulting firm, MARTEC Ltd., for the Tidal Power Corporation, the project proponent.[114] Nova Scotia's Minister of the Environment, after review of the document by the Department's Assessment Division, presented recommendations to the Tidal Power Corporation on conditions necessary for approval to proceed with construction.

In its EIS, MARTEC identified areas where data were inadequate or insufficient and indicated further studies critical to make a more informed assessment of predicted impacts. MARTEC, however, failed to discuss possible impacts from construction or operation on commercial or sports fisheries, focusing instead on fish stock status. The Assessment Division, in its review, concluded that questions remained about spoil disposal, scour, fisheries, water level management, basin hydraulics, bank erosion, groundwater, and agriculture. After agreement was reached between NS DOE and the Tidal Power Corporation on additional studies, the project was given conditional approval. Construction commenced in May 1980 and was completed in 1984.

The Intergovernmental Advisory Committee for the Annapolis Tidal Power Project was established by NS DOE to review the studies which the Tidal Power Corporation agreed to conduct as well as to monitor the project during its construction. Late in 1983, the secretary of the Intergovernmental Environmental Advisory Committee reported that uncertainty remained in five areas: agricultural monitoring, fisheries resources, water quality, flow forecasting, and flood routing. Studies in these areas continue.

The Tidal Power Corporation's handling of the EIS resulted in some controversy. Before the MARTEC report was released for public review, the chairman of the TPC announced a call for tenders stating that this process need not be delayed until public review of the EIS. While public protest remained peripheral in the assessment phase, it provided the stimulus for the Corporation to establish a compensation fund to cover losses resulting from project construction and operation.

Conclusion

Although there are dangers involved in generalizations based on only a few examples, the case studies considered here nonetheless share suggestive similarities. The

EARP process, despite its flexibility and its direct, early infusion of outside expertise, is often ineffective in counteracting the narrowing effect of initial decisions. Detailed procedures, however, do not counter the effects of narrowing as is demonstrated by application of the U.S.'s NEPA.

The force and quality of these early decisions indicate that evaluation of indirect, far-field impacts suffer particularly because of their inherently elusive nature. Legislating their consideration apparently does not improve their chances of being considered. Rather only a changed way of thinking about assessment, one that focuses on specifying unknowns, uncertainties, and disagreements, will ensure scrutiny for these issues. Such a change best comes about as a conscious decision on the part of the assessors.

Some of the delays and lack of coordination that characterized decisionmaking for the Pittston project would be unlikely if a similar project were proposed today. The protracted permitting for Pittston stimulated a more streamlined approach to federal permitting in the U.S.[115] In an attempt to ward off the litigation stemming from the project and others involving resource conflicts, interagency review has been strengthened. Thus, procedural reforms now contribute to a more efficient process.

Less clear is the degree to which improvements have been achieved in decisionmaking concerning substantive issues. Agencies tend to cling to their initial assumptions and so a flawed analysis of substantive issues may remain difficult to correct through outside agency or public review. Proposed resource use projects now receive unprecedented scrutiny by the public, environmental groups, and the scientific community; however, decisionmaking concerning exploratory oil and gas leasing on Georges Bank,[116] for example, exhibited similar agency attempts to buttress early, faulty analyses using incomplete data rather than to modify the evaluation. The difficulty seems to arise from assessing impacts whose effects are difficult to predict. Here information is often lacking, of poor quality, and subject to disagreement by experts.

Because indirect, far-field, transboundary effects commonly exhibit these qualities, specific regulations requiring their consideration may be of little utility despite extensive review. What is necessary is for evaluators to go beyond the requirements. Informal exchanges of information among experts in an atmosphere where they are not burdened by agency or national affiliation represents one way to move beyond meeting the letter of the law. These exchanges provide a unique opportunity to point out inadequacies in analyses early in decisionmaking. Finally, assessments must contain an evaluation of the adequacy of information, indicating areas of disagreement, data gaps, and uncertainties.

Participation of affected parties beyond political boundaries represents one clear way to ensure that far-field transboundary impacts are included in the assessment process. Such inclusion, however, trespasses on the sensitive area of sovereignty. Neither Canada's nor the U.S.'s formal requirements for the consideration of environmental effects outside territorial boundaries specifies review and comment from those bearing the consequences of the proposed action. Assessment in Canada, though, does show a trend in this direction. U.S. citizens were invited to participate in hearings for Canada's proposed Beaufort Sea exploratory oil and gas drilling program and Denmark and Greenland participated in a similar hearing for the Arctic Pilot project.[117] If the assessment process is viewed as an information-gathering activity upon which decisions are based, then it must guarantee the participation of all those potentially affected.

Notes

1. Federal Environmental Assessment Office (FEARO) 1985. Environmental Assessment in Canada - 1985 Summary of Current Practice.

2. *Ibid.* p. 37.

3. FEARO 1984. Environmental Assessment and Review Process Guidelines Order, SOR/84-467.

4. Beanlands, G. and P.N. Duinker. 1983. An Ecological Framework for Environmental Impact Assessment in Canada. Institute for Resource and Environmental Studies, Dalhousie University and Environment Canada.

5. U.S. National Environmental Policy Act, 42 U.S.C. §§ 4321-4361.

6. Council on Environmental Quality (CEQ). NEPA Regulations. 40 CFR 1500, 43 FR 55990 (November 29, 1978).

7. *Ibid.*, § 1502.

8. *Ibid.*, § 1502.22(b).

9. Council on Environmental Quality Proposal to Revise NEPA Regulations to Remove Requirement for Worst-Case Analysis. 40 CFR 1502, 50 FR 32234 (August 9, 1985).

10. *Ibid.*, proposed 1502.22(d), 50 FR at p. 32238.

11. *Ibid.*

12. Rosenbaum, K.L. 1985. Amending CEQ's Worst Case Analysis Rule: Towards Better Decisionmaking? Envt'l Law Reporter, Vol. 15, pp. 10275-10278.

13. Bureau of National Affairs 1986. Env't Reporter Current Developments, pp. 1812-13 (Jan 31, 1986).

14. *Supra* note 6, § 1508.8(b).

15. *Supra* note 6, § 1502.22(d).

16. Caldwell, L.K. 1982. Science and the National Environmental Policy Act. University of Alabama Press. p. 60.

17. Stewart, R.B. and J.E. Krier. 1978. Environmental Law and Policy. Indianapolis. The Bobbs-Merrill Company, Inc., pp. 803-808.

18. *Ibid.*

19. *Ibid.*

20. Caldwell, *supra* note 16 at 2.

21. *Ibid.* p 12.

22. Carroll, J.E. 1983. Environmental Diplomacy. Ann Arbor. University of Michigan Press. pp. 29-34.

23. Research for this analysis was largely conducted by John Roanowicz and forms the basis of Roanowicz, J. and J. Spiller. 1984. The Proposed Pittston Oil Refinery: Case Study of U.S. Environmental Decisionmaking. Working Paper -1. Marine Law Institute.

24. *Lehigh v. Pittston*, 456 A.2d 355 (Me. 1983).

25. *Roosevelt Campobello Park Comm'n v. U.S. E.P.A.*, 684 F.2d 1034, 1039 (1st Cir. 1982).

26. U.S. EPA. 1976. Draft Environmental Impact Statement: Proposed Issuance of Federal Permits to the Pittston Company for the Construction of a 250,000 Barrel/Day Oil Refinery and Marine Terminal -Eastport, Maine. Vol. I-II.

27. U.S. EPA. 1978. Final Environmental Impact Statement: Proposed Issuance of Federal Permits to the Pittston Company for the Construction of a 250,000 Barrel/Day Oil Refinery and Marine Terminal -Eastport, Maine. Vol. I-VI.

28. *Ibid.*

29. Letter from the Fish and Wildlife Service to William R. Adams, Region I U.S. EPA Administrator (December 21, 1978); Letter from National Marine Fisheries Service to Arnold F. Kaulakis, Vice-President for Energy Development, The Pittston Company (March 8, 1979).

30. *See Roosevelt Campobello v. EPA*, *supra* note 25, for a discussion of the Administrative Law Judge's opinion.

31. *Roosevelt Campobello v. EPA*, *supra* note 25.

32. Letter from J.R. McKinney, Canadian Embassy Minister in U.S. to R.Z. Smith, Dep'ty Asst. Sec'y for Canadian Affairs, U.S. State Dept. (December 6, 1972).

33. Personal communication M. Ludwig, Habitat Protection Division, NMFS, Northeast Fisheries Center (NEFC), Milford, Ct. with John Roanowicz, Complex Systems Research Center (CSRC), University of New Hampshire (March 1983).

34. Letter from R.D. Vine, Dep'ty Asst. Sec'y for Canadian Affairs to John A.S. McGlennon, Administrator, U.S. EPA, Region I (July 7, 1976).

35. Letter, *supra* note 33. In February of 1973, these pilots joined the Atlantic Pilotage Authority, a Canadian Crown corporation. Thus, they required formal government permission to testify.

36. Maine Board of Environmental Protection, Finding of Fact and Order (March 12, 1975).

37. Personal Communication Ludwig, *supra* note 33.

38. Scarratt, D.J. Ed. 1979. Evaluation of Recent Data Relative to Potential Oil Spills in the Passamaquoddy Area. Canadian Fisheries and Marine Service Tech. Report 901.

39. *Roosevelt Campobello v. EPA*, *supra* note 25.

40. Scarratt, *supra* note 38; Trites, R. Ed. 1974. Summary of the Physical, Biological, Socio-economic and Other Factors Relevant to Potential Oil Spills in the Passamaquoddy Bay Region of the Bay of Fundy. Canadian Fisheries and Marine Service Tech. Report 28.

41. Canada Shipping Act. Oil Carriage Limitation Regulations. P.C. 1982-499. SOR/82-244.

42. Personal Communication with W. Stickney, Environmental Advisor, State of New Hampshire with John Roanowicz (CSRC) (March 1983).

43. Clean Water Act 33 USC § 1344; Endangered Species Act 16 USC 1531 *et seq.*

44. Hughes, S. 1979. A Case Study of the Pittston Company's Proposal for Construction of an Oil Refinery and Marine Terminal at Eastport, Maine. In *Energy Development Project Delays: Six Case Studies*. Congressional Research Service Rept. Serial 96-7.

45. Endangered Species Act Final Regulations. 43 FR 873 (January 4, 1978).

46. Rosenberg, R.H. 1981. Federal protection of unique environmental interests: Endangered and threatened species. *Land Use and Environment*. Vol. 12 pp. 469, 486.

47. 42 FR 4868 (Jan. 26, 1977).

48. *NWF v. Coleman*, 529 F.2d 359 (5th Cir. 1976), *cert. denied* 429 U.S. 579; *TVA v. Hill*, 549 F.2d 1064 (6th Cir., 1977), *aff'd* 437 U.S. 153 (1978).

49. Hughes, *supra* note 44.

50. U.S. EPA, *supra* note 26 at 129; U.S. EPA, *supra* note 27 at 25.

51. Scarratt, *supra* note 38; U.S. EPA, *supra* note 27 at 25.

52. U.S. EPA, *supra* note 27 at 56-4.

53. Memo on July 24 trip through Head Harbor Passage from W. Stickney, Environmental Impact Office, U.S. EPA, Region I to J.A.S. McGlennon, Regional Administrator, U.S. EPA Region I (July 28, 1975).

54. Enviro-Sciences, Inc. 1976. An Environmental Assessment Report for a Proposed 250,000 BPD Fuel Refinery and Deepwater Marine Terminal at Eastport, Maine, U.S.A.; U.S. EPA, *supra* note 26.

55. *Ibid.*

56. U.S. EPA, *supra* note 27.

57. *Ibid.*

58. U.S. EPA, *supra* note 27.

59. *Ibid.*

60. Trites, *supra* note 40.

61. Spaulding, J.L. 1979. An Analysis of the Behavior of a Hypothetical Oil Spill in the Bay of Fundy near Campobello Island. Prepared for the National Oceanic and and Atmospheric Administration and the New England Fishery Management Council.

62. U.S. EPA, *supra* note 27.

63. Sanders, P. 1977. Wind Stress on the Ocean over the Eastern Continental Shelf of North America. *Jour. Phys. Oceanography*. Vol. 7, pp. 555-556.

64. Lauzier, L.M. 1967. Bottom Residual Drift on the Continental Shelf Area of the Canadian Atlantic Coast. *Jour. Fish. Res. Bd. Canada*. Vol. 24, pp. 1845-1859.

65. Trites, *supra* note 40.

66. U.S. EPA, *supra* note 27.

67. Letter from W.A. O'Neil, Dep'ty Administrator of Marine Services, Canadian Department of Transport, to A.F. Kaulakis, Vice-President for Energy Development, The Pittston Company (March 31, 1973).

68. Canada Coast Guard. 1977. Eastport Ship Terminal System: Accessibility and Ship Safety: Preliminary Analysis and Assessment. Can. Coast Guard, Ottawa; Department of External Affairs. 1976. Eastport Ship Terminal System: Accessibility and Ship Safety, Preliminary Analysis and Assessment. *Eastport*; Fisheries and Environment. 1976. An Environmental Risk Index for the Siting of Deep Water Oil Ports. Department of Fisheries and Environment; Trites, *supra* note 41.

69. Letter U.S. Coast Guard to U.S. EPA Region I (March 28, 1977).

70. U.S. EPA, *supra* note 27.

71. *Roosevelt Campobello v. EPA*, *supra* note 25.

72. U.S. EPA, *supra* note 26, at 129; U.S. EPA, *supra* note 27 at I-25.

73. *Ibid*. p. I-25.

74. *Ibid*. p. VI-39.

75. Comments of S.K. Katonah. College of the Atlantic to Environmental Impact Statement Group, U.S. EPA Region I (Dec. 22, 1976).

76. *Ibid*.

77. U.S. EPA, *supra* note 27 at 2-1. NMFS comments to U.S. EPA on DEIS.

78. Internal memo from M. Ludwig, Habitat Protection Division, NMFS, NEFC, Milford, Ct. Endangered Species Chronology and Pittston (Feb. 2, 1979).

79. Letter from W. Stickney, Environmental Impact Division, U.S. EPA Region I, to W. Gordon, NMFS, NEFC (June 23, 1978).

80. Beanlands and Duinker, *supra* note 4.

81. King, J.E. and J.G. Nelson. 1983. Evaluating the Federal Environmental Assessment and Review Process with Special Reference to South Davis Strait, Northeastern Canada. *Environmental Conservation*. Vol. 10, pp. 293-301.

82. Schrecker, T.F. 1984. Political Economy of Environmental Hazards. Law Reform Commission of Canada.

83. Beanlands and Duinker, *supra* note 3.

84. *Ibid.* p. 106.

85. *Ibid.* p. 106.

86. Imperial Oil Limited, Aquitane Company of Canada, Ltd. and Canada Cities Services, Ltd. 1978. Environmental Impact Statement for Exploratory Drilling in South Davis Strait Region. Calgary, Alberta. 384 pp.; Imperial Oil Limited, Aquitane Company of Canada, Ltd. and Canada Cities Service Ltd. 1979. Supplement: Environmental Impact Statement for Exploratory Drilling in South Davis Strait Region.

87. Beanlands and Duinker, *supra* note 4 at 106; King and Nelson, *supra* note 81 at 297.

88. Beanlands and Duinker, *supra* note 4 at 106 citing FEARO. 1978. Report of the Environmental Assessment Panel, Eastern Arctic Offshore Drilling - South Davis Strait Project. Ministry of Supply and Services. Ottawa, Ontario. 80 pp.

89. *Ibid.*

90. Beanlands and Duinker, *supra* note 4 at 106.

91. Eastern Arctic Marine Environmental Studies (EAMES) Advisory Board. 1979. Report of the EAMES Advisory Board on Exploratory Drilling in South Davis Strait Region.

92. King and Nelson, *supra* note 81 at 296, citing Department of Fisheries and Oceans. 1978. Review of the Environmental Impact Statement for Exploratory Drilling in the Davis Strait. Regional Hydrocarbon Committee 7.

93. Beanlands and Duinker, *supra* note 4 at 107.

94. *Ibid.*

95. *Ibid.*

96. *Ibid.* p. 109.

97. *Ibid.* p. 110.

98. *Ibid.* p. 106, citing Fisheries and Oceans, *supra* note 92.

99. Beanlands and Duinker, *supra* note 4 at 108, citing Department of the Environment 1978. Review of Environmental Impact Statement for Exploratory Drilling in the Davis Strait. Hydrocarbon Committee. Edmonton, Alberta. p. 6.

100. EAMES, *supra* note 91.

101. King and Nelson, *supra* 81 at 296, citing Department of the Environment, *supra* note 99.

102. *Ibid.* p. 7.

103. FEARO, *supra* note 88 at 35.

104. Department of Energy, Mines and Resources. 1978. Review of the Environmental Impact Statement: Exploratory Drilling South Davis Strait. p. 19.

105. FEARO, *supra* note 88 at 16.

106. *Ibid.* pp. 41-42.

107. King and Nelson, *supra* note 81 at 296.

108. *Ibid.*

109. *Ibid.* p. 297.

110. Bay of Fundy Tidal Review Board. 1977. Reassessment of Fundy Tidal Power. Wolfville, Nova Scotia.

111. Federal Environmental Assessment and Review Office. 1977. Guidelines for Fundy Tidal Power.

112. *For example*, Gordon, D.C. and M.J. Dadswell, Eds. 1984. Update on the Marine Environmental Consequences of Tidal Power Development in the Upper Reaches of the Bay of Fundy. Canadian Tech. Rept. Fish. Aquatic Sciences -1256.

113. *For example*, Greenberg, D.A. 1979. A Numerical Model Investigation of Tidal Phenomena in the Bay of Fundy and Gulf of Maine. Marine Geodesy Vol. 2.

114. MARTEC. 1980. Environmental Assessment Report of the Proposed Tidal Power Development on the Annapolis River System.

115. Hughes, *supra* note 44.

116. *Conservation Law Foundation v. Clark*, 586 F.Supp. 1238 (D. Mass. 1984).

117. Bruchet, D. 1983. Commens. In M.I.T. Sea Grant. *Arctic Technology and Policy.*

CHAPTER 5

Seeking Legal Redress for Transboundary Environmental Injury

Robert A. Taylor

Introduction

One option that the United States and Canada may utilize to resolve transboundary environmental problems may be described as the remedial approach. The term 'remedial' suggests that rights have been violated, that some level of transboundary damage has already occurred, and that efforts are directed toward assigning compensation responsibility to redress the damage. In addition to compensatory remedies, injunctions may be sought to prevent a violation of rights that has not yet transpired but is threatened. Underlying both categories of remedies is the preventive function that adequate *post hoc* remedial actions may serve. Where compensatory legal remedies for transboundary damages are perceived as available and effective, persons or organizations undertaking activities with possible transboundary impacts will be aware of their potential liability and can be expected to take steps to reduce the likelihood of damage.

Discussions of remedial approaches must distinguish between remedies of an essentially private nature and those of a public nature. Activities in one country that inflict damage in another country may be grounds for a private remedy through the domestic legal system of either country. The same damage may also be considered a violation of an international obligation owed by the nation within whose jurisdiction the damage-causing activity takes place to the nation within whose jurisdiction the damage occurs. This violation of rights on an international level may be grounds for a public remedy, through governmental resort to negotiation, arbitration, international judicial processes, or other mechanisms. Different doctrines and theories of liability and judicial jurisdiction are applicable depending on whether remedies are being sought on the private, transnational plane or on the public, international plane. Therefore, these two classes of remedies will be considered separately.

Private Remedies

Attempts by private citizens to obtain judicial remedies for transboundary environmental damage caused by large-scale energy developments could result in lawsuits involving several possible parties and courts. In the case of damage in the

United States caused by the operation of the Fundy tidal power facility, U.S. citizens could bring action against Canadian defendants in either United States or Canadian courts. In the case of damage in Canada caused by the U.S. tidal power development, Canadian citizens could bring action against U.S. defendants in either Canadian or United States courts.

The United States and Canada have similar legal systems and share many doctrines and theories of jurisdiction, procedures, and remedies applicable to the type of lawsuits under consideration. On several issues, similar questions may arise regardless of the nationality of the parties or court. Where legal theories in the two countries have diverged, the choice of court may result in substantial differences in the adequacy and availability of legal remedies. This section briefly describes some of the hurdles that would have to be overcome in actions for transboundary environmental damage in both Canadian and U.S. courts, also pointing out difficulties unique to either country's legal system. The intent here is not to be exhaustive, but merely to show by selected example the range of problems that could be encountered.

Jurisdiction and Immunities

Private litigants may face certain jurisdictional limitations in transboundary environmental disputes either in the form of judicially-created doctrines or in the terms of statutes that provide bases for recovery. One such limitation that would be applicable to suits against an entity such as the Fundy Tidal Power Corporation is the doctrine of sovereign immunity.[1] Sovereign immunity, the idea that the king (or government) can do no wrong and is therefore not subject to suit, has traditionally barred suits against foreign as well as domestic governments. Immunity has also been extended to the actions of governmental agencies and instrumentalities such as government-owned corporations like the Tidal Power Corporation.

The 1976 Foreign Sovereign Immunities Act (FSIA)[2] was enacted to regulate the practice of U.S. courts relating to sovereign immunity, and, in certain circumstances, to create exceptions to the immunity rule - that is, to permit suits to be brought. One of the primary exceptions relevant here is in the case of "an act outside the territory of the United States in connection with a commercial activity of the foreign state elsewhere and that act causes a direct effect in the United States."[3]

A private action by an aggrieved U.S. citizen against a Canadian utility that is a Crown corporation would potentially come within this exception. Under the Act, "foreign state" includes a political subdivision and its agencies and instrumentalities, a definition that would apply to Crown corporations. The Act's legislative history, however, makes it unclear whether the creation of transboundary environmental damage would be considered a "direct effect," or the production of electricity a "commercial activity."

Even if the FSIA would preclude the utility from invoking immunity to bar damage actions against it, a related rule may be applied to achieve immunity. Under the act of state doctrine,[4] United States courts have refused to litigate the validity of actions by a foreign government within its own territory. The act of state doctrine has not been altered by legislation, and has been used by courts to avoid adjudication of cases that the FSIA would permit. Cases discussing the doctrine provide unclear guidance as to the circumstances that compel its application. It does appear that the act of state doctrine is more likely to be applied when the litigation in question involves sensitive foreign relations issues.[5] Thus, the doctrine's use in a Fundy tidal power lawsuit could well depend on how successful the Canadian government could be in making the case an international political issue.

Actions by citizens against the domestic government are also subject to the rule of sovereign immunity, a rule that U.S. citizens seeking remedies in Canadian courts against the tidal power operators would likely face. In the United States, federal and state statutes waive sovereign immunity in the case of certain tort actions. Similar statutes have been enacted in Canada. As with the FSIA, the availability of immunity in such cases becomes a matter of statutory interpretation.

Apart from the issue of immunity, U.S. citizens bringing actions in Canadian courts would probably be precluded from obtaining relief for damage to real property located in the United States. The local action[6] rule prevents courts from deciding cases involving injury to real property that is located outside the court's territorial jurisdiction. Many United States courts recognize an exception to the rule in cases where the damage-producing activity takes place within the court's jurisdiction.[7] Canadian courts have apparently not yet adopted this exception,[8] and would likely reject such an action by U.S. citizens.

Standing

Assuming that jurisdictional problems can be overcome, questions of standing to seek relief may have to be answered. The rules of standing require a plaintiff to satisfy the court that the plaintiff has a legally protected interest in the lawsuit. Standing to recover for property damage under common law theories is generally established by demonstrating that the plaintiff has lawful possession of the damaged property. When the right to recover is provided by statute, the plaintiff may be required to show that his or her interests are intended to be protected by the rights the statute creates.

Theories of Relief

Recovery for transboundary environmental damage may be based on several common law theories of relief. Theories that may be utilized include nuisance, trespass, negligence, strict liability, and the public trust doctrine. Recovery may be based on one or more of these theories, and court decisions often do not clearly distinguish between them.

Nuisance is defined as the unreasonable interference with the use and enjoyment of land. Thus, pollution or environmental modifications that damage or inhibit the use of real property may constitute a nuisance. Relief under a nuisance theory may depend upon whether or not the nuisance is deemed private rather than public. A public nuisance affects the population of a community or region, and generally may be legally enforced only by public officials.[9] Individual relief in the case of public nuisances will only be granted for injury that is different in kind from that which is suffered by the general public.[10] A Maine resident whose shoreline property is damaged by erosion or flooding attributable to the effects of a tidal power facility, for example, may be required to show that his or her injuries were different in kind from those of shorefront property owners throughout the Gulf of Maine region.

Nuisance actions may also be faced with the defense of statutory or regulatory authority[11] - that is, the idea that damage-producing activity conducted under governmental license or permit presumptively is not a nuisance. As in the case of the public/private distinction, court decisions interpreting this rule have limited its effect, at the same time making its application more uncertain.

An activity that results in a physical invasion of real property may give rise to a right to relief under a trespass theory.[12] While courts in recent years have tended to expand the applicability of trespass notions to permit relief for environmental damage, proving a physical invasion of property may still be an insurmountable hurdle in some cases.

Recovery for environmental damage may also be possible if the plaintiff can demonstrate that the defendant was negligent. Proof of negligence requires showing that the defendant's conduct violated a duty of care owed to the plaintiff, and that the conduct caused the plaintiff damage.[13] The rights that may be vindicated in a negligence action are not limited to land ownership and a negligence suit therefore may be more attractive than an action for nuisance or trespass in certain circumstances. Difficulties with negligence actions include proving causation and identifying the standard of care, a component of which is the foreseeability of damage.

Strict liability, often discussed in conjunction with negligence theories, does not involve the defendant's fault in the sense of the violation of a duty of care. Rather, strict liability will result when damage is caused by an activity that is abnormally dangerous.[14] Although case law establishes that certain activities are judicially considered abnormally dangerous,[15] for a unique technology that determination will turn on the facts of the particular case. Courts often focus on the location of the activity in question and the risks it creates in considering whether the danger presented is abnormal or unreasonable.

U.S. citizens bringing an action in U.S. courts may also seek relief under the public trust doctrine. Although the public trust doctrine is historically derived from British legal theories, Canadian courts have been slow to adopt it. The doctrine provides that certain natural resources, traditionally including submerged and public trust tidal lands and marine resources, are deemed held in trust by the government for the benefit of and common use by the public.[16] Where the government does not act to protect public trust resources, citizens may in some cases use the public trust doctrine to compel governmental action.

A private litigant may also try to use the public trust doctrine as a basis for relief for environmental damage. As in the public nuisance cases, similar questions may be raised as to whether the individual plaintiff has suffered damage unique from that suffered by the general public.[17] The doctrine has been used successfully in at least one case by commercial fishermen to recover for damage to fishery resources caused by an oil spill.[18] The court reasoned that the fishermen suffered unique damage by the invasion of the public right to fish. It is uncertain whether such reasoning would also permit shorefront property owners to recover for damage to uplands in connection with damage attributable to a tidal power facility to public trust tidal lands.

In addition to the affected private individuals, governments may also seek judicial remedies in domestic legal systems for transboundary environmental injury. Goverments within whose jurisdiction environmental injury occurs may, under the theories of public nuisance or public trust, institute litigation to obtain relief. Governments may also take advantage of the doctrine of *parens patriae*, which some courts recognize as granting governments a remedy for injury to natural resources.[19] *Parens patriae* is essentially a special rule of standing, permitting recovery where the government demonstrates that it has a sovereign interest in resources beyond that claimed by its citizens.

Problems of Proof

Regardless of which theory of recovery is used, establishing a right to relief for environmental damage may be most seriously limited by problems of proof. A plaintiff has the burden of proving entitlement to relief and that the defendant is liable under some theory. For the tidal power project, proof of causation may be especially problematical where damage is produced not by the addition of a substance or contaminant to the environment (such as oil), but by the alteration of natural phenomena such as current or tidal regimes or oceanographic parameters. For example, it may be impossible as a practical matter to prove that the alteration of the tidal regime by six or seven inches produced erosion or flooding damage greater than that which otherwise would have occurred. The lack of a scientific consensus on the type or degree of damage that changes in the tidal regime are likely to cause may make courts unwilling to find that plaintiffs have carried the burden of proof. Differences in scientific opinion may also affect proof of foreseeability of damage, affecting both the determination of the standard of care under a negligence theory and the characterization of an activity as abnormally dangerous under a strict liability theory.

Remedies

When the defendant's liability for injury has been established, the plaintiff is potentially entitled to either monetary compensation, an injunction, or both. Compensation, or 'damages,' is intended to restore the plaintiff to the economic position enjoyed before the physical damage occurred. Just as problems of proof may attend the assignment of liability for environmental injury, proof problems can also hinder the computation of damages. Damages may be based on estimates derived from existing data, but courts are unwilling to assess damages purely on speculation.

The granting of injunctive relief is even more problematical since injunctions are considered to be within the equitable powers of a court. In applying equitable remedies, courts are permitted to exercise great discretion, balancing the advantages and hardships to each party of granting or denying the injunction. In addition, courts will usually deny an injunction where an award of monetary damages is deemed adequate.

Enforcing Judgments

A plaintiff who has obtained a favorable judgment in a transboundary environmental damage case may face difficulties in enforcing the judgment. As a consequence of the rule of foreign sovereign immunity, the property in the United States of a foreign government is granted immunity from execution in satisfaction of judgments. The FSIA creates exceptions to the immunity rule in the case of property of a foreign government used for a commercial activity in the United States upon which the claim is based,[20] and property "of an agency or instrumentality of a foreign state engaged in commercial activity in the United States."[21] Assuming that the Canadian tidal power utility's dealings in the United States are limited to sales of electricity to U.S. utilities, both exceptions would require a finding that such sales constituted "commercial activity in the United States." In addition, the utility must, of course, have property within the United States.

If the tidal power operator has no property within the United States subject to execution, U.S. plaintiffs may seek to enforce the judgment in Canada. Canadian courts generally recognize and will enforce a foreign judgment if the judgment is founded on a theory of jurisdiction that would be applied in a Canadian court. Because U.S. courts are generally more liberal than Canadian courts on matters of long-arm jurisdiction, any uncertainties surrounding the assertion of jurisdiction by U.S. courts over a foreign defendant would be compounded in seeking foreign enforcement of a favorable judgment. Foreign enforcement of an injunction is highly questionable, given that the enforcing court would reweigh the equities of the injunction and may not agree that it is warranted.

A hybrid between private and public remedies is an international claims commission. Such a commission may be established by agreement between two countries to settle claims by the nationals of one country against the nationals or government of the other, usually arising out of some specific incident or dispute. The agreement creating the commission may contain provisions relating to jurisdiction, establishing a right to relief, and enforcing successful claims. Yet, to the extent that the agreement does not obviate the need to prove issues of causation and damages, such a commission would simply constitute an alternative forum rather than a more certain means of obtaining relief.

The uncertainties surrounding private judicial remedies affect all aspects of litigation, from the issues of obtaining jurisdiction and proving a right to relief, to obtaining satisfaction of a favorable judgment. Besides these difficulties, lawsuits are generally focused exclusively on the parties to the litigation and a specific source of conflict. Larger issues or a more comprehensive consideration of affected interests are rarely within the scope of judicial proceedings. An international claims commission may deal more effectively with the larger issues, but may present problems of proof as insurmountable as those encountered in domestic legal systems. Further, because of the low likelihood of the availability of effective injunctive relief, both private remedies and the quasi-private remedy of an international claims commission suffer from the drawback of focusing on redressing past damage rather than on preventing future injury. Private and quasi-private remedies, therefore, have little to offer as a future option for managing transboundary resources or avoiding future transboundary problems.

Public International Law

Public international law consists of the body of norms of international conduct, the obligations nations must observe in their activities affecting each other. Examination of whether certain activities constitute a violation of international law includes identifying the substance of the obligation, determining whether the violative conduct should be attributed to the nation under whose jurisdiction the conduct takes place, and considering international remedial processes that may be used to redress the violation.

The Standard of Liability

International law authorities are in general agreement that a nation is under an obligation not to inflict environmental damage on persons or property in other nations. The seminal statement of this obligation in its modern form is found in the 1941 opinion of the *Trail Smelter* arbitral tribunal.[22] The tribunal was established by the United States and Canada to resolve disputes surrounding the operations of an ore smelter at Trail, British Columbia, fumes from which were claimed to have caused property damage in the State of Washington. Applying analogous principles of international law and domestic law of the United States, the tribunal concluded:

> no State has the right to use or permit the use of territory in such a manner as to cause injury by fumes in or to the territory of another or the properties or persons therein, when the case is of serious consequence and the injury is established by clear and convincing evidence.[23]

Although the tribunal's holding is directed on its face to transboundary transmission of fumes, the reasoning behind the opinion applies equally well to other environmental impacts. Other international judicial opinions, examples of national practice, and statements of principles adopted by international bodies support the principle of *Trail Smelter* and its application to transboundary environmental damage in general.[24]

Some authorities argue for a different formulation of the obligation concerning transboundary impacts. Drawing an analogy from theories in the law relating to international watercourses, several writers contend that a principle of 'equitable use' applies in all cases involving the use of an internationally shared natural resource.[25] The doctrine of 'equitable use' bases the international legality of a nation's activities affecting a shared resource on whether those activities deny other nations sharing the resource their opportunity to make equitable use of the resource. 'Equitable use' requires a balancing of the benefits and detriments of each nation's use of the resource to determine the most equitable division of the resource's capabilities. Extension of this theory to any shared resource would make it applicable to an area such as the Bay of Fundy-Gulf of Maine region whose waters are partially enclosed by the coastlines of the United States and Canada.

The primary distinction between the *Trail Smelter* principle and the doctrine of equitable use is that *Trail Smelter* focuses largely upon impacts within the damage-exposed nation, while an 'equitable use' analysis considers equally the ramifications of the relevant activity for the damage-exposed and the damage-producing nation. In actual application, the two theories would likely lead to the same result since an activity which produces transboundary damage substantial enough to violate the *Trail Smelter* rule would probably not be considered an 'equitable use' of the shared resource.

Attribution of Responsibility

Public international law is mainly concerned with the conduct of nations rather than with the conduct of individual citizens. Yet a nation can only act through the conduct of individuals or groups of people. Distinctions must therefore be drawn between those actions that should be considered purely private and those that should be attributed to the national government for international law purposes. In cases in which the substantive obligation that has been violated is one prohibiting the production of transboundary environmental damage, the question raised is whether the national government within whose jurisdiction the damage-producing activity takes place should be held internationally responsible for that activity and the resulting damage.

Most authorities are in agreement that simply proving that damage has occurred is not sufficient to establish international responsibility. Thus nations are not subject to a strict liability standard in transboundary damage cases.[26] Some showing must be made that the national government itself was at fault, that some activity attributable to the government led or contributed to the damage. When the damage-producing activity is carried out by an arm or subdivision of the government, attribution is rarely an issue since the government itself is acting to create the transboundary damage.

In private activity cases, for example if the damage occurs as a result of the operation of a privately-owned industrial facility, governmental involvement is usually less direct, instead involving the government as permitter or regulator. The argument may be made that the conduct attributable to the government is the regulation of the private activity, and that such conduct contributed to the transboundary damage by not being reasonably adequate to prevent the damage from occurring. The focus of the analysis thus shifts to the level of control that the government actually exercised and the level that it should have exercised under the circumstances.

An activity producing transboundary impacts, such as the proposed operation of a tidal power facility by a government-owned utility, would likely give rise to international liability on the part of Canada. Action by a corporation owned by an internal governmental subdivision would be considered conduct attributable to the national government for international law purposes. By comparison, the operation of a privately-owned oil terminal and refinery in Maine would not likely be conduct for which the United States government would be considered internationally accountable. Determination of whether any transboundary damage that resulted from the facility would give rise to the United States' liability would depend on whether the degree of regulatory control exercised by governmental authorities over the facility was reasonably adequate in light of the potential for damage.

Mechanisms for Redress

Assuming a case may be made that the relevant national government is responsible in some manner for the production of transboundary environmental damage at a level to violate either the *Trail Smelter* or 'equitable use' standard, several avenues are available to the injured nation in seeking redress. Disputes between nations may be resolved through a process of negotiation which may result in informal settlement arrangements or in a formal agreement. Negotiations may also lead to an agreement to submit the matter to a third-party dispute settlement mechanism such as international arbitration.

In certain circumstances nations may be subjected to third-party settlement involuntarily. The Statute of the International Court of Justice, the judicial arm of the United Nations, provides that nations may adopt declarations recognizing as compulsory the Court's jurisdiction over international disputes, subject to reservations.[27]

The United States has made reservations in its jurisdictional declaration that could be very important to Bay of Fundy-Gulf of Maine region problems. The United States' reservations include "disputes with regard to matters which are essentially within the domestic jurisdiction of the United States of America as determined by the United

States of America."[28] Such a reservation would permit the United States to remove any dispute from the Court's jurisdiction by unilaterally declaring it to be a matter within the United States' domestic jurisdiction. Canada had a major reservation to the jurisdiction of the International Court of Justice but recently withdrew the reservation. When Canada enacted the Arctic Waters Pollution Prevention Act in 1970, which imposed a special 100 nautical mile pollution prevention zone in arctic waters, Canada was concerned about being subjected to international litigation since the innovative legislation was likely beyond coastal state powers. Canada, therefore, withdrew from the compulsory jurisdiction of the International Court of Justice over law of the sea disputes. It removed from the Court's jursidiction "disputes arising out of or concerning jurisdiction or rights claimed or exercised by Canada in respect of ... the preservation or control of pollution or contamination of the marine environment in marine areas adjacent to the coast of Canada."[29] However, on September 10, 1985, External Affairs Minister Joe Clark, when announcing the drawing of straight baselines (effective January 1, 1985) around Canada's Arctic Archipelago, also declared Canada was withdrawing the former reservation since the law of the sea had evolved to include sufficient coastal state powers.

By virtue of the principle of reciprocity, in a case between the United States and Canada either nation could take advantage of a reservation to the World Court's jurisdiction the other had adopted. Thus in a case involving environmental damage in the Bay of Fundy - Gulf of Maine region, Canada might argue that the U.S. reservation places the matter outside the Court's compulsory jurisdiction.

The reservation of such disputes from compulsory jurisdiction may also affect legal measures that the Court could adopt to prevent threatened transboundary damage. The Court is authorized to "indicate ... provisional measures which ought to be taken to preserve the respective rights of either party."[30] In effect, the Court is granted the power to issue an interim injunction to protect a party pending the resolution of the case. Thus, for example, the United States might bring an action before the Court to prohibit further construction or operation of a tidal power project.

In previous cases, the Court has been faced with requests for interim protective measures that were opposed on the grounds that the dispute was subject to a reservation and thus outside the Court's jurisdiction. The Court's response to such arguments has been to consider itself empowered to indicate protective measures so long as the Court is satisfied that the dispute appears to fall within its jurisdiction.[31] However, the Court's determinations on jurisdiction in previous cases suggests that predictions on jurisdictional decisions in interim measures cases cannot be made with any degree of certainty.[32]

Unilateral Actions

Considering the uncertainty of judicial preventive measures, a nation may be tempted to take unilateral actions to prevent threatened environmental damage. Such actions may properly affect other nations or their citizens, short of violating international obligations. For example, nations are generally recognized to have broad authority to regulate the movement of people and goods across their frontiers. Thus, the United States would be within its rights internationally, absent conflicting treaty obligations, to restrict the importation of electricity from Fundy tidal power facilities that cause transboundary damage.

One type of border-crossing activity that traditionally has enjoyed substantial freedom from national regulation involves the movement of vessels through certain ocean areas. Although a ship moving through a nation's territorial sea crosses that nation's borders, international law recognizes limits upon the extent to which the coastal nation may regulate the ship's activities. Yet so long as those limits are not exceeded, a coastal nation may adopt unilateral regulatory measures to protect its interests.

Canada's adoption of regulations on tanker traffic through Head Harbor Passage is one example of such a unilateral measure. Titled the Oil Carriage Limitation Regulations, the rules were enacted in February, 1982, and provide in part: "No oil tanker that is in the waters of Head Harbor Passage, New Brunswick, shall carry on board, as cargo or otherwise, oil in excess of 5000 m^3."[33] The 5000 m^3 of oil permitted was an estimate of the needs of the Eastport, Maine region and would have prevented the passage of supertankers to and from the proposed Eastport refinery. Although Canada has asserted since 1973 the right to restrict tanker traffic in the area, the 1982 regulations were the first effort to do so.

Predicting the outcome of a dispute over the reasonableness of the regulations would be difficult. Canada has taken the position since 1970 that passage that creates a hazard of environmental damage is non-innocent, a position that the United States has opposed.[34] However, it seems fair to suggest that the uniquely treacherous navigational characteristics of Head Harbor Passage combined with the very defined reach of the restrictions would support the regulations' validity.

The disadvantages of the use of unilateral international measures to prevent transboundary environmental damage include the tendency of such measures to foster a confrontational attitude between affected parties. In addition, the potential scope of unilateral efforts is restricted and they often are unable to adequately deal with the ultimate source of the threatened transboundary damage. Yet, in limited circumstances they may in fact be quite effective in addressing particular problems. Further, unilateral measures may occasionally be politically attractive as a means of avoiding a difficult or weak negotiating position by relying instead on a principle of international law.

Notes

1. *See* Comment, "Act of State and Sovereign Immunity Doctrines: The Need to Establish Congruity," 17 U.S.F.L. Rev. 91, 93-96 (1982); *The Schooner Exchange v. McFadden*, 11 U.S. (7 Cranch) 116 (1812).

2. 28 U.S.C. §§1601-1611 (1976).

3. 28 U.S.C. §1605(a)(2).

4. *See generally* Comment, *supra* note 1.

5. E.g., IAM v. OPEC, 649 F.2d 1354, 1358 (9th Cir. 1981).

6. The case commonly cited for the local action rule is *British South Africa Co. v. Companhia de Mocambique*, [1983] A.C. 602. *See* McCaffrey, "Transboundary Pollution Injuries: Jurisdictional Considerations in Private Litigation Between Canada and the United States," 3 Cal. W. Int'l L.J. 191, 217-229 (1973).

7. *See, e.g.*, Armendiaz v. Stillman, 54 Tex. 623 (1881); *Ducktown Sulphur, Copper & Iron Co. v. Barnes*, 60 S.W. 593 (Tenn. 1900); *Vermont Valley R. Co. v. Connecticut River Power Co.*, 99 Vt. 397, 401, 133 A. 367, 371 (1926); *Mannville Co. v. Worcester*, 138 Mass. 89, 91 (1884); *cf. Arizona Commercial Mining Co. v. Iron Cap Copper Co.*, 119 Me. 213, 215 (1920) 110 A 429, 431.

8. *See, e.g., Albert v. Fraser Companies, Ltd.*, 1 D.L.R. 39, 11 Mar. Prov. 209 (N.B. 1936); *Brereton v. Canadian P.R. Co.*, 29 Ont. 57 (1899); *Winnipeg Oil v. Can. Nor. Ry.*, 18 W.L.R. 421, 21 Man. 274 (1911); *Boslund v. Abbotsford Lumber, Mining and Development Co. Ltd.*, 1 D.L.R. 978 (B.C. 1925).

9. W. Rodgers, *Environmental Law* 102 (1977).

10. *Id.* at 106; McCaffrey, "Private Remedies for Transfrontier Pollution Damage in Canada and the United States: A Comparative Survey," 19 W. Ont. L. Rev. 35, 47-48 (1981); 1 R. Franson & A. Lucas, *Canadian Environmental Law* 3.1.1.3 (1976) (Six-volume looseleaf). *See, e.g., Hickey v. Electric Reduction Co. of Can. Ltd.*, 21 D.L.R. 3d 368 (Nfld. 1972), and *Union Oil v. Oppen*, 501 F.2d 558 (9th Cir. 1974).

11. W. Rodgers, *supra* note 9, at 136-42; R. Franson & A. Lucas, *supra* note 10, at 3.1.2.2; *e.g., New England Legal Foundation v. Costle*, 666 F.2d 30, 33 (2d Cir. 1981).

12. W. Rogers, *supra* note 9, at 154.

13. R. Franson & A. Lucas, *supra* note 10, at 3.3; McCaffrey, *supra* note 10, at 51.

14. W. Rodgers, *supra* note 9, at 158.

15. *Fletcher v. Rylands*, 3 H. & C. 774, 159 Eng. Rep. 737 (1865), *rev'd* 1 L.R.-Ex. 265 (1866), *aff'd*, 3 L.R.-E.&I. App. 330 (1868). *See also*, W. Rodgers, *supra* note 9, at 159.

16. W. Rodgers, *supra* note 9, at 170-71; McCaffrey, *supra* note 10, at 40-41.

17. McCaffrey, *supra* note 10, at 53.

18. *Burgess v. M/V Tamano*. 370 F. Supp. 247 (D.Me. 1973).

19. *See, e.g., Maine v. M/V Tamano*, 357 F. Supp. 1097 (D.Me. 1973).

20. 28 U.S.C. §1610(a)(2) (1976).

21. 28 U.S.C. ²1610(b) (1976).

22. Trail Smelter Case (U.S. v. Can.), 3 R. Int'l Arb. Awards 1907 (1941). *See generally* Rubin, Pollution by Analogy: The Trail Smelter Arbitration, 50 Or. L. Rev. 259 (1971); Read, *The Trail Smelter Dispute*, 1 Can. Y.B. Int'l L. 213 (1963); Dinwoodie, *The Politics of International Pollution Control: The Trail Smelter Case*, 27 Int'l J. 219.

23. *Id.* at 1965-66.

24. *E.g.*, Corfu Channel Case (U.K. v. Alb.), [1949] I.C.J. 4; Statement of Canadian Secretary of State for External Affairs, House of Commons Debates, June 8, 1972, at 2995, *reprinted in* 11 Can. Y.B. Int'l L. 333-34 (1973) (statements of Canadian authorities in connection with a 1972 oil spill in the State of Washington that fouled British Columbia beaches); Report of the United Nations Conference on the Human Environment, Stockholm, June 5-16, 1972, U.N. Doc. A/CONF.48/14 and Corr. 1, *reprinted in* 11 I.L.M. 1416 (1972); Organization for Economic Cooperation and Development: Council Recommendations on Principles Concerning Transfrontier Pollution, adopted Nov. 21, 1974, OECD Doc. C(74)224, *reprinted in* 14 I.L.M. 242 (1975).

25. *E.g.*, Handl, "The Principle of 'Equitable Use' As Applied to Internationally Shared Natural Resources: Its Role in Resolving Potential International Disputes Over Transfrontier Pollution," in Organization for Economic Cooperation and Development, *Transfrontier Pollution* 98 (1981).

26. *See.*, *e.g.*, Dupuy, "International Liability of State, for Damage Caused by Transfrontier Pollution," in Organization for Economic Cooperation and Development, *Legal Aspects of Transfrontier Pollution*, 345, 355 (1977); Handl, "Balancing of Interests and International Liability for the Pollution of International Watercourses: Principles of Law Revisited," 13 Can. Y.B. Int'l L. 156 (1975).

27. Charter of the United Nations and Statute of the International Court of Justice, Article 36 of the Statute, 59 Stat. 1031, 1060 T.S. No. 993.

28. Declaration on the Part of the United States, Aug. 14, 1946, 61 Stat. 1218, T.I.A.S. No. 1598, 1 U.N.T.S. 9.

29. Canadian Declaration Concerning the Compulsory Jurisdiction of the International Court of Justice, *reprinted in* 9 I.L.M. 598 (1970).

30. Statute of the International Court of Justice, *supra* note 9, Article 41, 59 Stat. at 1061.

31. *E.g.*, Nuclear Tests Case (Austl. v. Fr.), [1973] I.C.J. 99, 102 (Request for the Indication of Interim Measures of Protection).

32. *Id.*

33. Oil Carriage Limitation Regulations, P.C. 1982-449, SOR/82-244, 116 Can. Gaz. [Part II] 892 (Oct. 3, 1982).

34. Department of State Statement on Government of Canada's Bills on Limits of the Territorial Sea, Fisheries and Pollution, 52 Dep't St. Bull. 610 (1970), *reprinted in* 9 I.L.M. 605 (1970); Summary of Canadian Note of April 16 Tabled by the Secretary of State for External Affairs in the House April 17, *reprinted in* 9 I.L.M. 607 (1970).

Chapter 6

Decisionmaking Improvements and Alternatives

David VanderZwaag, Gordon Beanlands, Peter Duinker, Karen A. Massey, Alison Rieser, Judith Spiller, Robert A. Taylor, and Peter Underwood

Introduction

Academics who suggest improvements or alternatives to decisionmaking frameworks operate under the shadow of a practical reality. Environmental decisionmaking is a dynamic *process* involving complex interrelationships among public interest groups, scientists, project proponents, and government officials who advocate value positions at various decisionmaking levels -- legislative and regulatory drafting, budgetary policymaking, federal permit processes, provincial or state licensing reviews, contract negotiations, and judicial proceedings. Thus, no suggested legal or administrative changes can guarantee smooth operation of decisionmaking in practice. Much depends on personal perceptions and a willingness of all parties to cooperate in reaching a resolution.

Because of these practical constraints, this chapter does not attempt to construct a model of "the perfect" or "the right" way to reach decisions. Rather it suggests ways of strengthening the Canadian and American decisionmaking processes and examines bi-national options for facilitating cooperative decisionmaking concerning tidal power development and other ocean uses with transboundary implications for the Bay of Fundy-Gulf of Maine region.

The first section of this chapter suggests modifications to the Canadian environmental assessment review processes to better ensure all interested parties receive a fair hearing on value positions, and to facilitate the collection and use of scientific information. Considerable attention is given to the Canadian process because it bears the major responsibility for reviewing Fundy tidal power's environmental effects. The second section addresses the need to clarify the United States legal framework for reviewing transboundary effects of Canadian energy projects like Fundy tidal power. The final section briefly explores alternative bi-national approaches for facilitating cooperative ocean management in the Bay of Fundy - Gulf of Maine region including informal, *ad hoc* consultations, reference to the International Joint Commission (IJC), cooperation at the state-provincial level, creation of new formal mechanisms such as a Fundy Tidal Power Commission, and adaptation of the United Nations Regional Action Plan approach for managing shared marine environments.

Improving the Canadian Environmental Decisionmaking Framework

Most of the procedures in Canada for acquiring and reviewing environmental information were not established with a view to the scientific credibility of the efforts involved. This is because such procedures were conceived and put into effect by bureaucrats responding to politicians under pressure from the general public to demonstrate an environmental ethic. Practicing scientists seldom get involved in bureaucratic exercises so there was little scientific perspective in such developments. Furthermore, since the term "environment" covers such a broad range of physical, biological, and social sciences, the scientific and technical implications are almost beyond definition. In effect, administrative procedures are out of balance with the ability of the scientists to respond to the demands as perceived by the general public.

As a result, and as demonstrated elsewhere in this review, there is a relatively poor record of the incorporation of scientific information in the decisionmaking process. There are, however, some changes which could be or are being implemented which may improve the application of scientific information to the resolution of environmental problems.

More Direction for Scientific Studies

A number of critical reviews of impact assessment procedures in Canada have underlined the need to provide a better focus for the component scientific studies. Environmental impact statements are characterized by a very broad approach to potential environmental problems associated with development proposals, with little evidence that they reflect priority concerns. With the limited time and resources normally available for impact assessments, it is crucial that more directed and focused studies be undertaken. This might be accomplished in a number of ways.

Early Identification of Major Issues. The term "scoping" has become associated with environmental assessment, meaning the identification of the major concerns associated with a particular development proposal. Under the revised National Environmental Policy Act (NEPA) regulations it is a required activity for all major assessments in the U.S., however, it is not yet a standard practice in Canada.

The idea is to have interested parties express their perceptions and concerns about a project proposal in order to provide some basis for determining priority issues involved. The issues thus identified become the major focus of attention during the impact assessment studies and reviews. While it has been attempted in some indirect ways in Canadian assessments, it has only recently been recognized as a requirement involving specific activities.

The environmental assessment conducted for the proposed second nuclear reactor at the Point Lepreau generating station in southern New Brunswick included scoping activities. The federal and provincial governments appointed members to an Environmental Assessment Review Process (EARP) panel which ran scoping workshops, where members of the general public were invited to express their concerns. These concerns then formed the basis for more specific guidelines to the proponent for undertaking environmental studies. Depending upon the outcome of this trial run, scoping procedures may be incorporated into EARP procedures with a consequent tightening up of the assessment guidelines and resulting scientific studies.

"Fast Track" Assessments. - Consideration is being given to having an impact assessment initially examine the most critical aspects of a proposed project with regard to key decision requirements. The objective is to obtain an early agreement in principle, following which more detailed studies would be undertaken to specify design changes, mitigation measures, and compensation.

Such a fast track approach has obvious time advantages for the proponents, and perhaps for politicians who may want to make some early commitments. However, it is unclear whether agreements in principle to proceed with a project should be made when more detailed assessment may reveal consequential environmental effects which were not obvious from a more superficial examination. It would seem that the fast track approach has a potential for undermining the credibility of environmental assessment as an important contribution to the decisionmaking process since the most fundamental decision is whether or not to proceed with the proposed project.

Clear Separation of Science and Values. - One of the major challenges facing decisionmakers involved with environmental problems is the separation of human values from scientific facts. Controversy has revolved around the definition of "acceptable." From a scientific perspective, information is considered unacceptable if it does not reflect the application of scientific criteria. However, unacceptable in a social context is a matter of personal values and judgment, i.e., the information may be acceptable scientifically but the implications from the perspective of the general public may be unacceptable.

In some cases, EARP panels conduct technical hearings (such as with the Alaska Highway Gas Pipeline) separate from general public sessions. This is an effort to determine if the technical information base of the impact statement is valid, before addressing the public acceptability of the changes which are forecast. A scientific advisory committee was established as part of the impact assessment for the second nuclear reactor at Point Lepreau. This committee, made up of recognized scientific experts from a number of disciplines, was formed to work with the proponent in designing a scientifically sound approach to the study of major project concerns. This early cooperation between government scientists and the proponent is a departure from the more normal adversarial approach adopted for environmental assessment in Canada. The hope is that the EARP panel will be able to focus its attention on the public acceptability of the predicted changes rather than adjudicate between arguing scientists during the public hearings.

More Involvement of the Scientific Community

In many cases, the problem of incorporating scientific information into the environmental decisionmaking process relates to the quality of the science involved. This is a two-fold problem. On one hand, studies of poor quality, involving questionable assumptions and few data, make it difficult for government panels or boards to draw meaningful conclusions. Under these circumstances, the normal procedure is to ask for supplementary information or clarification before the study reports are considered acceptable and public hearings can proceed.

On the other hand, when such studies are critiqued by government scientists during normal review procedures, their negative comments often throw the credibility of the entire study program into question. Such criticisms, when made public, make it difficult for the review body to trust the content of environmental assessment studies.

Many of these problems relate to the reluctance of the scientific community to become involved in environmental impact assessments. One of the difficulties is that the performance criteria for government research scientists does not include involvement in assessment studies. Since such involvement does not normally lead to scientific publications in recognized journals, they cannot afford the time and effort since it goes unrecognized in their career evaluations. This problem can only be solved through changes in performance criteria as part of personnel management of government departments.

Perhaps even more fundamental to the problem of the lack of scientific involvement is the question of scientific acceptability of the research involved. The Fundy Environmental Studies Committee is an example of the range and quality of environmental research which can be undertaken in support of major development proposals if the interest of the scientific community can be attracted and maintained. It is far more common, however, for the scientists to consider assessment studies as short-lived, mission-oriented activities designed to support decisions already taken.

The question of a research base for environmental impact assessment lies at the heart of the problem. Only if impact assessment is seen to be a challenging area for scientific research will it be possible to draw in good scientists. In this regard, the recent establishment of the Canadian Environmental Assessment Research Council (CEARC), through the initiative of the Federal Environmental Assessment Review Office (FEARO), should provide a focus for such activities. CEARC has the mandate to promote long-term research in the physical, biological, and social sciences as it relates to improving the scientific and technical basis for environmental impact assessment and related activities such as baseline studies and monitoring. It is the intention to have CEARC recognized as a credible and permanent "champion" for high-quality assessment research which will ultimately influence the research budgets of major government departments.

More Attention to Assessing Development Scenarios

Those responsible for making decisions with regard to large developmental proposals, such as the Fundy tidal power project, often become overwhelmed with tedious details. When the effects of a complicated development are superimposed on a multitude of physical and social environments, the result is often beyond the comprehension of many individuals.

As demonstrated by the recent Beaufort Sea impact assessment, review panels are requesting information on alternative broad development scenarios. These scenarios, when considered in a more general way, make it easier for decisionmakers to grasp the total implications of a project and weigh alternative development strategies. In responding to such requests, proponents are making more use of simulation modeling with the consequent reliance on sophisticated analytical systems. In some ways, the complexity of the simulation modeling adds to the problems of the decisionmakers since they require technical advice if they get involved with the mechanics of the scenario building. However, in the Beaufort Sea case the evaluation of various development scenarios seemed to be the only way for the assessment panel to focus on the most important decision options.

It is likely that the use of development scenarios, as opposed to an amalgamation of individual project impacts, will be used more often to assist decisionmakers pondering major environmental problems. This may have particular relevance to the Fundy project since Canadian scientists are currently working on adopting an estuary simulation model developed in the Netherlands for predicting the effects of a tidal barrage in the Bay of Fundy. The successful development of such a model would presumably permit the effects of various project configurations to be assessed.

More Coordinated Follow-up to Environmental Assessments

One of the most crucial deficiencies in environmental assessment as conducted in Canada is the lack of procedures to ensure that panel recommendations are carried out and to monitor the actual effects of the project. As a result, panels have no way to assess clearly the effects of previous, similar projects. Each panel begins its deliberations with the assumption that the project in hand is totally unique, and the assessment guidelines are drafted accordingly.

The assessment conducted for the proposed second nuclear generating unit at Point Lepreau is an example of how helpful follow-up monitoring can be. The assessment panel for the first nuclear unit recommended that a comprehensive biological monitoring program be established. As a result, in the mid-1970's the federal Department of Fisheries and Oceans instituted a scientific monitoring program covering both marine and terrestrial environments. As a result of this program, the panel considering the second nuclear unit had the results of both pre-operational baseline studies and post-startup monitoring at its disposal. This will be one of the few examples in Canada where a monitoring program has been able to assess the cumulative effects of a phased development. Future projects will also be subject to long-term monitoring studies, for Section 33.1(d) of the Environmental Assessment and Review Process Guidelines Order requires the initiating federal department to ensure that "suitable implementation, inspection and environmental monitoring programs are established."

Clarification of the Substantive Base of Environmental Decisionmaking

The present EARP guidelines and other federal and provincial permitting processes leave great discretion to decisionmakers as to the extent environmental values and scientific information are to be weighed in reaching a final decision. For example, the EARP Guidelines grant almost total discretion to the initiating department in choosing to implement environmental conditions:

> [I]t is the responsibility of the initiating department in a public review to ... decide in cooperation with any other department, agency or board of the Government of Canada to whom the recommendations of a Panel are directed, the extent to which the recommendations should become a requirement of the Government of Canada prior to authorizing the commencement of a proposal.[1]

To clarify the role of science and ecological "fact" in decisionmaking, legislative criteria should be established providing guidance to decisionmakers as to the extent ecological impacts predicted by scientists should be weighed against other values - social, economic, technological, and political. Various substantive standards are possible, each bestowing a potentially different level of environmental protection. For example, a cost-benefit approach might be stipulated. If the environmental and social costs of a proposed project outweigh the social and economic benefits, then the decisionmaker might be mandated to reject the project.[2] A risk-benefit approach might also be invoked which bids a decisionmaker to weigh the probability (or risk) of environmental injury and the potential gravity of harm against the socio-economic benefits of a project and the technological costs of environmental protection.[3] A significant environmental impact standard might also be stipulated as a legal touchstone for decisionmaking. If a project is deemed to have a significant adverse environmental impact (which might be defined in more specific ways such as any loss in recruitment of an endangered species), the decisionmaker might be required to disapprove the project.

The formulation of substantive criteria should press decisionmakers towards greater "objectivity" through a requirement to articulate clearly the value basis of judgments, but in the end some subjectivity seems inevitable. The selection of substantive criteria (for example, that the decisionmaker is to give priority to renewable resource harvesting over nonrenewable resource uses) is itself a subjective choice. The weighing of unquantifiable factors, such as the risk of long term environmental effects or the potential reduction of intertidal zones or salt marshes, also involves a subjective balancing. Nevertheless, decisionmakers should be subject to the value choices in substantive criteria imposed by elected officials through the legislative process.

Strengthening the Legal Bases of Environmental Assessments

At the federal level, the EARP process, although substantially strengthened by the Environmental Assessment and Review Process Guidelines Order of June 21, 1984, still operates with at least two major weaknesses. First, the process still suffers from legal uncertainties. Do the Cabinet's Guidelines impose a process with binding legal obligations reviewable and enforceable by a court? Or are the Guidelines indeed guidelines which are still subject to substantial governmental discretion? To remove all doubt as to the legal status, the EARP process should be given a firm legislative base. Second, the EARP process leaves final decisionmaking largely with the initiating federal department. To assure the Minister of Environment has sufficient bargaining strength to represent environmental interests, the Minister should be granted actual decisionmaking authority for federal projects which may cause a significant adverse environmental impact. Because a broad expansion of ministerial power at the potential expense of the power of other departments may be unrealistic, the power to override ministerial decisions would likely remain with the federal Cabinet.

At the provincial level, Nova Scotia and New Brunswick both have rather minimal requirements for environmental assessments which could be substantially strengthened. Nova Scotia not only operates a highly discretionary environmental assessment process without written guidelines available to the public, but also provides no guaranteed public input into the ministerial decisions to grant or refuse permits required by the Environmental Protection Act and the Water Act. The Environmental Protection Act does allow the Environmental Control Council to hold public hearings in relation to plan approvals, permits, and orders made by the Minister, but the Minister of Environment would have to authorize such hearings.[4] To provide a fair opportunity for all interests to participate in the decisionmaking process and to assure all major projects are subject to adequate environmental assessment, the provincial assessment process should be established in detailed legislation. Public notice should be required of project proposals before issuance of ministerial permits and public reivew should be mandatory when there is an expression of public concern or when the environmental impacts may be significant according to established criteria.

In New Brunswick the environmental assessment process presently operates under the modest authority of a Cabinet policy adopted on October 8, 1975, but a draft regulation (Environmental Impact Assessment Regulation - Clean Environment Act) is being reviewed which could substantially strengthen the process. As presently drafted (December 2, 1985), the regulation would require public review of environmental assessment reports - public review is presently discretionary - and would list undertakings likely to result in significant environmental impacts and normally requiring a formal environmental assessment (for example, electric power generating facilities of three megawatts or more, water reservoirs, and electric power transmission lines exceeding 69 KV capacity or exceeding five kilometers in length).

A number of additional strengthenings, however, should be considered. Even though Section 31.1(2)(n) of the Clean Environment Act allows the Lieutenant-Governor in Council to pass regulations imposing criteria for approving or disapproving projects after completion of an environmental impact assessment, the draft regulation provides no such guidance to the Minister of Municipal Affairs and Environment, and Cabinet would retain final approval authority. The draft regulation also fails to guarantee public participation at the early stages of the review process, including the formulation of guidelines and the terms of reference. Public meetings are only assured after actual preparation of an environmental impact assessment report. Clear substantive criteria for decisionmaking should be established and full public participation should be assured even at the initial assessment stages.

Coordinating Federal-Provincial Reviews

Present legislation leaves open the potential for fragmented and duplicative environmental reviews. The provinces might require extensive studies as precondition for issuance of land-use or water-use permits, which federal reviews (not necessarily in the public arena) might be undertaken by the Departments of Fisheries and Oceans, Environment, Transport, and the National Energy Board.

Federal-provincial duplication would likely be avoided or moderated through a cooperative, project-specific environmenal assessment. Both Nova Scotia and New Brunswick have entered into environmental accords (whose status often seems uncertain) with the Federal Department of the Environment and have pledged to cooperate in environmental assessments.[5]

However, since numerous marine-related projects, such as Fundy tidal power, will involve jurisdictional overlaps of provincial property, resource management rights, and powers over social and economic conditions, with federal control over fisheries, navigation, and international relations, the two levels of government should consider creating a more formal process for undertaking joint environmental assessments. Instead of following a project-by-project, *ad hoc* approach, the governments should implement a continuing, long-term environmental planning process established by identical federal and provincial legislation.[6] A permanent federal-provincial Coastal Zone Environmental Assessment and Planning Commission or Board might be formed in each Atlantic province and the Commission would operate with clearly defined mandates and procedures, including:

1. A requirement to consider all potential adverse impacts of proposed projects, including socio-economic impacts, transboundary implications, and indirect and cumulative impacts on fisheries and wildlife;

2. A requirement to consider technological alternatives to proposed project designs;

3. A requirement to involve the public at the earliest stages of environmental review;

4. A requirement to develop in cooperation with other governmental departments and representatives from the public an ocean use management plan (or regional plans) identifying present ocean uses, potential conflicts, and methods for avoiding future conflicts; and

5. A delineation of substantive criteria for reaching decisions.

A more formal process could emphasize the need to involve all interest groups - affected individuals, scientists, government officials, and proponents - during the entire process of decisionmaking. The full involvement of relevant interest groups is essential to focus scientific inquiries and to ensure process-oriented hypothesis testing.

The formation of such federal-provincial coastal zone commissions might be part of a larger program of federal-provincial cooperation in water management and development. The recent Inquiry on Federal Water Policy recommended that to facilitate cooperation in freshwater management the federal government should develop master agreements with the provinces as part of a National Water Conservation Program. The agreements would provide for: 1) a general commitment to cooperate in water matters and set out the principles and conditions of cooperation; 2) supplementary agreements on specific collaborative programs; and 3) consultative arrangements to oversee joint undertakings.[7] Coordinated assessment in coastal management could follow this model.

Improving the United States' Environmental Decisionmaking Framework

Individual national action is not the optimum approach to resolving transboundary pollution problems, but the United States could take some internal actions that would improve environmental assessment of energy project development in transboundary regions such as the Gulf of Maine.

In the United States, clarification is needed of the decisionmaking framework for review of transboundary effects of Canadian energy projects, where there is a physical connection with the United States via transmission lines. There are several options for providing this clarification: broad interpretation of the National Environmental Policy Act (NEPA), amendment of the Department of Energy (DOE) rules, amendment of the executive order governing the presidential permit for transboundary transmission lines, amendment of existing federal legislation, and amendment of state legislation.

Broad Interpretation of the National Environmental Policy Act

One option is to pursue the theory that the National Environmental Policy Act, properly interpreted, requires any environmental impact statement (EIS) prepared in connection with the presidential permit for a transboundary power transmission line also to address the impacts in the United States from the Canadian generating facility.

In at least one instance NEPA was so interpreted. When General Public Utilities Corporation (GPU) sought approval for a cable under Lake Erie and for accompanying surface lines to connect Ontario-Hydro's power generation with the Pennsylvania system, the EIS prepared by the Department of Energy considered the potential air quality impacts in the U.S. from burning more high sulfur coal at the Nanticoke, Ontario coal-fired generating station, sixty kilometers from the U.S. border. While this precedent provides support for the view that transboundary impacts of the generating facility itself should be examined, it by no means assures that this would be the case. The discussion of transboundary air impacts consumed less than two pages of the GPU transmission line EIS. An adequate discussion of the impacts of the proposed Fundy tidal power facility would have to dwarf any discussion of the transmission line impacts.

Political considerations may affect whether the U.S. Department of Energy would choose to discuss Fundy tidal power's effects in an EIS, but whether these considerations would favor or disfavor such an interpretation of NEPA could vary depending on the context at the time the EIS is prepared. Were such a discussion not included in the draft EIS distributed for public review, interested groups or members of the public could submit comments on the draft EIS and, if necessary, engage in court action to argue the theory that such impacts are within the proper scope of an EIS. Although the potential environmental impacts of Fundy tidal power are much more complex than in the GPU case, the principle is the same: construction of transmission lines primarily or solely to import power from a particular Canadian plant should consider transboundary impacts of the plant itself. Such impacts could be the basis for denying a permit--as not in the public interest--or for imposing mitigating conditions on the granting of a permit.

Amendment of Department of Energy Rules

Another route for clarification of this issue would be to petition DOE to amend its rules governing presidential permit applications to explicitly provide that environmental effects in the United States of related generating facilities, whether located here or abroad, must be considered in the EIS for the transboundary lines.

Amendment of the Executive Order

A third alternative would be to seek to influence the President to amend the executive order itself, rather than the DOE rules, governing the presidential permit process. The order could be amended in one of two ways: 1) to require specifically that impacts of the Fundy tidal power dam on United States territory be considered in review of the permit application, or 2) to require that transboundary environmental impacts of energy facilities generally be considered in presidential permits. Success in such an endeavor depends, of course, on a receptive, environmentally sensitive Executive Branch, and some means of access to present the case.

Amendment of Federal Energy Legislation

Procedures and channels for proposing legislative amendments are more established and accessible to citizens or citizen groups than are those for amending executive orders. Again, it would be possible to take either a general or a particular approach to legislative amendment. A general approach would amend that section of the Federal Power Act requiring approval of electricity exports to give the Department of Energy authority to review imports as well. This would parallel the Natural Gas Act, which regulates both exports and imports of gas.

A new subsection to the Federal Power Act could be added, stating:

> No person shall transmit or enter into agreement for the transmission of any electric energy into the United States from a foreign country without first having secured an order of the Secretary of Energy authorizing it to do so. The Secretary shall issue such order upon application, after opportunity for public hearing, and upon compliance with the requirements of the National Environmental Policy Act of 1969, if he finds that the proposed transmission is consistent with the public interest. The Secretary may by order grant such an application in whole or in part, with such modifications and upon such terms and conditions as the Secretary may find necessary or appropriate, and may from time to time, after opportunity for hearing and for good cause shown, make such supplemental orders as he may find necessary or appropriate.

This proposed amendment has three major facets. First, it provides for review of the proposed import prior to entry into a final contract (prior to construction in the case of a new facility), that is, at the preliminary agreement stage, to assure review at the earliest possible stage in the process. Second, it establishes a general public interest standard of review. (The standard of review for exports is simply whether "the proposed transmission would impede or tend to impede the coordination in the public interest of facilities subject to the jurisdiction of the ... [DOE].") Third, it explicitly provides for compliance with NEPA.

This legislative proposal may be criticized by some on the grounds that the general public interest standard opens the door to the incorporation of domestic protectionist policies or other factors that might not have been intended by, and could possibly interfere with the goals of, the drafters. Whether Midwestern utilities seeking to limit Canadian competition could be successful is one question this raises. Another question, outside the scope of this discussion, but being considered at the federal level, is whether, given the dramatic increase in U.S. electricity imports in recent years, the U.S. should not have a national policy on electricity imports. The objection that such a policy would not necessarily further the goals of controlling transboundary pollution might be dealt with by modifying the legislative proposal to delete the general public interest standard and more closely track the export standard quoted above, while still including language to insure that environmental considerations play a role in decisionmaking.

Another option is passage of special legislation specifically requiring NEPA review of Fundy tidal power impacts. This legislation would perhaps have a better chance of passage, as affected states would all favor it, and others would have little reason to oppose it. There is indirect precedent for such legislation in statutes *exempting* specific projects from NEPA consideration.[8] Without an import license requirement, however, the EIS might still be limited to a consideration of the transmission facilities.

Amendment of State Legislation

Because of the admittedly regional nature of Fundy tidal power impacts, another possibility would be to seek amendment of Maine state legislation to assure consideration of transboundary environmental effects. Again, the power line would be the primary trigger to involvement, although here approval of the contract could be the mechanism for control. Thus one might amend the state Public Utilities Commission's contract review authority to allow consideration of environmental effects, or broaden the existing Site Location of Development Act to encompass consideration of impacts from the generating facility. In this latter case, it would be more difficult to interpret existing state law as requiring such review than it would be to contend that the federal NEPA covers such "indirect" effects of transmission line permitting. The argument can be made, however, that the issue of transboundary effects should be dealt with on the federal level as a matter of foreign relations.

There is some truth in saying that there is regulation and probably a regulatory and policy gap that the United States could and should take steps to fill with respect to consideration of all the effects on the U.S. environment from the import of electricity. Because of the transmission line connection, the U.S. has a jurisdictional mechanism for addressing this issue through one of the channels outlined above.

Such unilateral action is subject to limitations however, such as Canadian refusal to provide information. The Pittston oil refinery case discussed in chapter 4 suggests that under some circumstances information useful to U.S. decisionmaking might be withheld. Canada might also resort to actions such as restructuring electricity export opportunities, if it objected to the U.S. policy.

These limitations of unilateral action point to the need for mechanisms for assuring that the U.S. and Canada cooperate in the control of transboundary pollution. More specifically, the countries must cooperate in the management of shared natural resources where actions on one side of the boundary produce effects on an ecosystem that ignores geo-political boundaries. While each nation should attend to improvements within its domestic decisionmaking process to assure preservation of environmental values, the sharing of information, cooperation in development of data bases, and finally cooperative decisionmaking seem to provide the best opportunity for optimal resolution of controversies involving transboundary regions.

Bi-national Decisionmaking Options

To facilitate cooperative environmental management, particularly as related to tidal power development in the Gulf of Maine - Bay of Fundy region, the U.S. and Canada may follow at least five major paths: informal, *ad hoc* consultations; reference to the International Joint Commission; cooperation at the state-provincial level; creation of new formal mechanisms such as a Fundy Tidal Power Commission; and the adaptation of United Nations Regional Action Plan approach.

Informal Consultations

The two countries could follow either the "high road" or the "low road" in informal consultations and cooperation. "High level" consultations could occur on at least four fronts: Canadian Prime Minister - U.S. President meetings, Ambassador - Executive/Legislative officials' meetings, Ministerial - Secretarial meetings (for example, between the Canadian Minister of Energy, Mines and Resources and the U.S. Secretary of Energy), and Canadian Parliamentarians - U.S. Congressmen meetings which occur formally through the Canada - U.S. Interparliamentary Group.

Lower echelon consultations and cooperation could occur on several major fronts. Review agencies in each country, such as the National Energy Board, Environmental Assessment Panels, and the Federal Energy Regulatory Commission, could open their doors to submissions by foreign residents and foreign officials. This route was taken when the National Energy Board allowed Greenland officials to intervene in Board hearings concerning the Arctic Pilot Project, a proposal by a Petro-Canada-led consortium to ship liquified natural gas from the Arctic to an eastern Canadian port. Similarly, the Beaufort EARP Panel, designated to examine the environmental issues surrounding hydrocarbon production in the Beaufort Sea, visited Alaska in 1981 for consultations with the North Slope Borough, industry, federal and state representatives, and academicians.[9] The panel subsequently allowed both U.S. and Danish/Greenland interventions.[10]

Moreover, the industrial proponent itself might cooperate with foreign government officials in carrying out environmental studies and in modifying industrial plans. For example, Petro-Canada formed a working committee with the Ministry of Greenland to study environmental concerns related to the Arctic Pilot Project. As a result of Greenlanders' concern Petro-Canada agreed to vary the proposed shipping route.[11]

Government officials may enter into rather informal arrangements for transboundary cooperation. For example, the Regional Director of the U.S. National Marine

Fisheries Service and the Director-General of the Scotia-Fundy region (Canadian Department of Fisheries and Oceans) have adopted consistent approaches to fisheries management in the transboundary region including equivalent meat counts for scallop harvests on Georges Bank and reciprocal spawning area closures for haddock on Georges Bank.

Academic and scientific institutions can facilitate transborder consultations. For example, in October, 1982, the Woods Hole Oceanographic Institution and the Dalhousie Ocean Studies Programme hosted a forum at Digby, Nova Scotia where government officials and academics from the United States and Canada discussed ocean management of the Fundy-Gulf of Maine region, including national approaches to fisheries management and marine environmental issues raised by Fundy tidal power. In January, 1985, the Canadian-American Centre of the University of Maine hosted a conference attended by both Canadian and American politicians, officials, and academicians, on the topic of free trade.

Scientific Cooperation

Perhaps the most significant cooperation could occur among scientific investigators. Recognition that the Bay of Fundy and Gulf of Maine form an interrelated ecosystem would represent a major contribution to improved decisionmaking in this region. Scientific studies reviewed in Chapter 2 indicate that the Bay and Gulf form an integrated system. Modeling the tides, in particular, has pointed out the unity of these waters. The driving physical forces influencing the tides also control the distribution of the biota.

The analysis of the use of science in decisions involving transboundary resources presented in chapter 4 indicates some of the pitfalls decisionmakers encounter when they ignore the interdependence of an ecosystem's components. One simple remedy for focusing analysis on too small an area is joint research and regular exchanges of information among researchers from many disciplines and from both sides of the boundary.

Several efforts in this direction have been made. These include joint research efforts between Canadian fisheries scientists at the Bedford Institute of Oceanography and the St. Andrews Biological Station and U.S. scientists at the Northeast Fisheries Center in Woods Hole. Additional projects include other institutions in the region.

Despite these activities, many research projects in the region continue to proceed unilaterally. In part, funding practices in both countries limit joint research. Since most funding comes from government sources, government attention in this area could contribute to expanding the scope of joint investigations. Funding mechanisms with some regional autonomy, for example in the U.S. Sea Grant program, could also stress joint efforts.

The Fundy Environmental Studies Committee, the cooperative endeavor, described in previous chapters, of researchers from Canadian universities and provincial and federal agencies studying the Fundy region, serves as a model for sharing scientific findings, developing joint research, and setting research priorities. The Committee has invited American researchers to participate.[12] In 1985, the New England Estuarine Society met in tandem with the Fundy group and should continue to do so. More informal exchanges are also useful. For example, physical oceanographers from Canada and the U.S. conducting research on Georges Bank and the Gulf of Maine meet every several years.

These joint activities have little relevance to important policy issues in the region unless those responsible for decisions are aware of the scope of research and pertinent findings. Scientists bear a special obligation to make public officials aware of the limits of scientific understanding, distinguishing between areas where more research will clarify issues and where additional information will make little contribution to resolving uncertainty. Decisionmakers familiar with the scientific and policy issues involved are better equipped to draw this crucial information from experts. Finally, evaluating farfield or indirect effects is a difficult process, involving a large component of judgment and hence a degree of risk on the part of the decisionmaker. That judgment is improved by direct contact with the scientific issues at question.

The International Joint Commission

Scientists and policymakers from both sides of the border already have a mechanism for collaboration on resource questions through the International Joint Commission (IJC). In an effort to resolve disputes over the use of waters forming and crossing the United States - Canadian boundary, the United States and Great Britain (acting on behalf of Canada) in 1909 adopted the Boundary Waters Treaty.[13] The treaty established the IJC, composed of six members, three each from the U.S. and Canada,[14] who are given regulatory jurisdiction over uses and activities that may affect the levels or flows of boundary waters. The treaty specified a series of conditions, requirements, and priorities for the exercise of this authority, known as the Commission's "applications" jurisdiction.[15] The term "boundary waters," those waters subject to IJC jurisdiction, is defined in the treaty.[16] Although the definition does not explicitly exclude marine waters, its wording suggests that boundary waters are limited to fresh water bodies. Nothing in the practice of the two countries indicates a belief that activities in marine waters would be subject to Commission approval. Thus, as presently worded and understood, the treaty would not provide for Commission regulation of energy developments in marine waters in the Bay of Fundy - Gulf of Maine system.

A related provision of the treaty prohibits "pollution" of boundary waters or waters crossing the boundary that would cause injury to persons or property on the other side of the boundary.[17] This prohibition also does not refer to marine waters and for similar reasons appears to encompass only pollution of fresh waters forming or crossing the boundary. The Commission is also provided with arbitral jurisdiction by the treaty, which may extend to practically any matter or difference between the two countries that they jointly decide to submit to the Commission.[18] This aspect of Commission jurisdiction has never been used.

In addition, the treaty gives the Commission a third type of authority in the nature of investigations and recommendations.[19] Called "reference" jurisdiction, this authority also covers any matter or difference arising along the common border, with the primary distinction between arbitral and reference jurisdiction being the non-binding nature of Commission reference reports. The invocation of the Commission's reference jurisdiction is completely voluntary on the part of the two governments.

Although either is permitted by the treaty to initiate a reference, the two countries in practice have referred matters to the Commission only on a joint basis. References have been conducted on a variety of issues, including the apportionment of boundary-crossing rivers (which are not subject to the Commission's applications jurisdiction), transboundary air and water quality problems, the fisheries impacts and economics of a joint United States-Canadian tidal power project in Passamaquoddy Bay, and the sociological problems of a small U.S. community in the State of Washington due to its location in relation to the international boundary. The specific question or questions to be examined in a reference will depend in each case on the nature of the controversy and the wording of the bilateral reference request.

Despite the variety of matters which a reference may cover, references generally have common elements and are conducted in a similar fashion. A reference usually involves the development of a joint body of information and technical data from which the Commission makes policy recommendations to the two governments. To perform the research and studies necessary to develop the information base, the Commission appoints a joint study board made up of experts from both countries. The study board members are usually governmental employees on leave or 'seconded' to the Commision, although while serving on the board they are expected to owe their primary loyalty to the Commission. The study board thus acts in a collegial, as opposed to a bilateral, fashion. That the joint study board mechanism can be effective in developing an agreed information base is evidenced by the fact that study boards have divided along national lines in only two of the 49 references conducted by the Commission.[20] In addition, several references have also led to the adoption of bilateral agreements designed to implement Commission recommendations and to provide a mechanism for preventing future problems of a similar nature.[21]

Transboundary environmental problems associated with Fundy tidal power development could be the subject of a Commission reference. This is clear from the variety of matters referred to the Commission in the past and from the Commission's conduct of a previous tidal power reference.[22] In that reference, the Commission reported in 1961 that the proposed joint project would disrupt fisheries in the immediate vicinity of the development although the impact on fisheries further away from the project would be minimal. However, the Commission's economic analysis showed that the power produced by the project would not be competitive with power from other sources and recommended against construction of the facility, a recommendation the two governments adopted.

Senator George Mitchell of Maine in 1983 requested the U.S. Secretary of State to begin discussions with Canadian officials to produce a joint reference to the IJC concerning the Fundy tidal power proposal. In a draft reference, the Senator suggested that the Commission draw upon both governmental and nongovernmental expertise from the two countries to determine the environmental and economic effects in the U.S. and Canada and to make a recommendation regarding development of the facility. While a reference on Bay of Fundy - Gulf of Maine transboundary problems may be appropriate, several limitations in the reference process suggest that it would not necessarily resolve the underlying dispute. A reference, as mentioned, can be effective in developing a jointly-agreed information base. The elimination of differences over factual matters can in turn reduce the dispute to policy questions, issues more familiar to diplomats and governmental decisionmakers.

However, it is in the area of policy considerations that the reference process has exhibited serious limitations. Issues requiring the making of policy decisions have been referred to the Commission by the governments without any agreed guidelines. The Commission thus finds itself in the position of making important policy choices for the two countries with little guidance as to the limits of political acceptability within which those choices must fall to be implemented. At the same time that the Commission is given too broad a scope on policy questions, its deliberations also are often too constricted in the sense that the Commission is usually required to make its policy decisions without reference to the larger political context of the bilateral relationship. As a result, the Commission may recommend policy choices which either or both governments are unwilling to implement because of the impact on other areas of the bilateral relationship outside the scope of the reference. These difficulties with the reference process are neither structural nor irreparable, but rather arise more from the practice of the two governments in framing the terms of reference.

The most obvious limitation to the reference process, of course, is its voluntary nature. Due to the practice of only submitting references bilaterally, the opposition of either government can prevent the initiation of a reference. When a matter referred to the Commission involves an activity already under way, delays in submission can preclude the Commission from recommending actions to prevent environmental damage that has already occurred. Further, either government may reject the Commission's recommendations in part or entirely, leaving the matter unresolved.

Several writers have suggested modifications in the Commission's authority or practice designed to increase the certainty of Commission involvement in transboundary environmental disputes and of implementation of Commission recommendations.[23] Those suggestions take the form of treaty amendments and interpretive protocols as well as changes in the staffing and structure of the Commission. Among the suggestions are expansion of the Commission's authority to give it regulatory jurisdiction over activities with transboundary environmental impacts, or to permit it to conduct references on its own initiative. In addition, separate agreements adopting international air and water quality standards could be concluded to provide guidance to the Commission in environmental references. However, any expansion of Commission authority must come at the expense of authority now exercised by existing governmental bodies such as the Departments of State and External Affairs. Writers who have considered the question argue that many of the suggested modifications in the Commission's role are politically infeasible.[24]

Perhaps a more politically acceptable alternative for increasing the likelihood of Commission involvement in Bay of Fundy - Gulf of Maine region matters would be for the two governments to submit to the Commission a standing reference on transboundary environmental problems in the region. This would permit the Commission to play a watchdog role, pointing out potential problems to the two governments before they become political disputes. Provision in the reference of a fixed budget earmarked for the Commission's regional role would permit the Commission to perform preliminary 'scoping' investigations of potential problems, enabling it to alert the governments to problems requiring deeper study and to indicate the form that study should take. The specification of a fixed term for such a reference would provide the two countries a formal means and opportunity to review its effectiveness and desirability, and would allay fears of the creation of a permanent international bureaucracy. Such a standing reference would have the advantage of not altering the basic relationship between the Commission and the governments, thus maintaining the political status quo while increasing the potential effectiveness of the Commission in handling transboundary environmental problems.

State-Provincial Cooperation

State-provincial cooperation in environmental protection is not an unusual occurrence in U.S.-Canadian regulations. For example, a New Brunswick-Maine agreement of June 28, 1973 called for close cooperation in the areas of environment, energy, trade, tourism, and transportation. Similarly, a British Columbia - State of Washington agreement pledged establishment of a legislative committee to coordinate pollution control regimes, particularly for oil pollution from tankers.[25] The eight states and two provinces near the Great Lakes have drawn up a Great Lakes Charter which pledges a united approach toward any future proposals for significant interbasin diversions of water.

For Eastern Canada and the United States, a special forum - the New England Governors' and Eastern Canadian Premiers' Conference - was created to facilitate bilateral consultations, as described in chapter 1. In the thirteen annual conferences held since the first conference in August, 1973 at Brundenell, Prince Edward Island, the forum has often considered environmental and energy questions. In 1979 the Conference resolved to support the timely conclusion of major electrical projects including hydro-electric developments in Quebec and Newfoundland and Fundy tidal power in the Maritimes. In 1980 the Conference passed a resolution renewing regional determination to accelerate development and exchange of hydro-electric energy. At the 11th conference in 1983 it passed a resolution calling for a reduction in acid rain with a target of no more than 20 kilograms of wet sulphate per hectare per year and also calling for the development of a mutually acceptable Transboundary Accord.[26] In June, 1984, at the 12th conference held at Newport, Rhode Island, the Conference established a Standing Committee on the Environment for promoting the exchange of scientific information, coordinating policies, strategies and control plans, and for reporting annually on the status of environmental issues including acid rain.[27] At the 13th annual conference in June, 1985, the Governors and Premiers passed a resolution encouraging the regulatory approval of the cross-border transmission of Nova Scotia natural gas.[28] It was also agreed that a letter would be sent to the Canadian Minister for External Affairs and the U.S. Secretary of State requesting that the Bay of Fundy tidal power project and its associated transboundary environmental issues be referred to the International Joint Commission.[29] The letter was subsequently sent by the co-chairpersons, Richard Hatfield, Premier of New Brunswick and John Sununu, Governor of New Hampshire, on June 18, 1985, but no reference to the IJC has yet occurred.

New Formal Arrangements/Policy Questions

Formal bilateral cooperation for ocean management in the Fundy-Gulf of Maine region is extremely limited. The U.S. and Canadian Coast Guards have developed a Joint Marine Pollution Contingency Plan which establishes the chain of command for responding to pollution events having transboundary implication and which apportions potential clean-up costs.[30] Chart planning in border waters is coordinated by a United States - Canada Hydrographic Commission. The Canadian Hydrographic Service has cooperated with U.S. hydrographers in carrying out ocean surveys in the Gulf of Maine region and five new charts are cooperatively being prepared for the boundary waters between Grand Manan Island and Passamaquoddy Bay.[31]

If the U.S. and Canada choose to enter new formal arrangements they will face several major policy questions. First, should they enter a demonstrative treaty (a general treaty expressing minimal intentions to cooperate) or a more ambitious administrative treaty (a specific treaty establishing concrete cooperative arrangements such as a joint ocean management or Fundy Tidal Power Commission)?[32]

A Fundy Tidal Power Commission, with equal U.S. and Canadian representation, might be established to undertake a cooperative environmental assessment of Fundy tidal power and to provide advice on the development of an adequate compensation scheme. The agreement could be modeled after the Canada-Denmark Marine Environment Cooperation Agreement, a primarily demonstrative treaty signed on August 26, 1983 to address marine environmental issues between Canada and Greenland.[33] The Agreement established a number of general commitments including consultation on projects with transboundary pollution potential, cooperation in exchanging scientific and regulatory information, and cooperation in identifying vessel routing areas outside territorial waters.

A demonstrative treaty would have two advantages. It would likely be politically acceptable as no great national commitments would be required by either country. It would give differing management regimes time to adjust to one another. On the negative side, a demonstrative treaty could encourage abundant international discussion but little concrete action.

Assuming the countries choose to enter an administrative treaty, should they create a new bi-national management institution(s) or should they simply agree to revise domestic legislation to facilitate international cooperation through existing domestic institutions? Revisions of domestic legislation might include such measures as:

a. Guaranteeing non-nationals the right to sue for environmental damages;

b. Requiring domestic environmental review entities such as the National Energy Board, Environmental Assessment Panels, and the Federal Energy Regulatory Commission to give notice of proceedings to interested foreign residents and relevant foreign governmental departments and to allow intervention by any such parties;

c. Allowing environmental review entities to fund foreign interventions and to perhaps even hold public hearings in another country, if agreeable with both governments;

d. Expanding general research into transboundary environmental issues; and

e. Harmonizing existing marine environmental approaches, for example, assuring consistent regulation of offshore petroleum operations.

The right of access to private remedies and administrative tribunals was recommended by a Joint Working Group of the American and Canadian Bar Associations in 1979. The Group suggested the following treaty provision:

Article 2: Rights of Persons Affected

a. The Country of origin shall ensure that any natural or legal person resident in the exposed Country, who has suffered transfrontier pollution damage or is exposed to a risk of transfrontier pollution, shall at least receive equivalent treatment to that afforded in the Country of origin, in cases of domestic pollution or the risk thereof and in comparable circumstances, to persons of equivalent condition or status resident in the Country of origin.

b. From a procedural standpoint, this treatment shall include but shall not be limited to the right to take part in, or have resort to, all administrative and judicial procedures existing within the Country of origin, in order to prevent domestic pollution, to have it abated, and/or to obtain compensation for the damage caused.

c. In the case of requirements for security of cost, this treatment shall at least be equivalent to that accorded to a non-resident national of the Country of origin.[34]

As a follow-up to the draft treaty suggestion, a liaison group of the National Conference of Commissioners on Uniform State Laws and the Uniform Law Conference of Canada, the two organizations in the U.S. and Canada dedicated to promoting uniformity of laws, drafted a model Uniform Transboundary Pollution Reciprocal Access Act.[35] The proposed legislation would provide reciprocal access to courts and administrative agencies of states and provinces which enact the legislation and become part of the "club".[36]

A partial precedent for an "internationalizing" of domestic reviews may be seen in both the U.S. and Canadian Clean Air Acts.[37] The Canadian Act authorizes the Minister of Environment to recommend air-emission standards for sources which "may reasonably be expected to constitute a significant danger to the health, safety or welfare of persons in a country other than Canada."[38] Cabinet would then have to establish procedures whereby the other country could make representation concerning the proposed standards.[39] The U.S. Act authorizes the Administrator of EPA to take action when he or she believes, on the basis of a report by a duly constituted international agency, that air pollutants originating in the U.S. "may reasonably be anticipated to endanger public health or welfare" in a foreign country. The Administrator may make a finding that will trigger a revision in state air quality plans to reduce or eliminate the transboundary pollutant. The affected foreign country must be invited to attend any public hearings convened on such revisions if recriprocal rights are available to the U.S. (42 U.S.C. 7415 (1982)).

A third policy question which arises is, assuming the U.S. and Canada agree to create a new bi-national management regime, how should they design their joint institution(s)? On the one extreme, they could create a multifunctional management commission, that is, a "super commission" with authority over all ocean uses, including renewable energy such as Fundy tidal power, fisheries, navigation, and oil and gas development. On the other extreme, they could establish one or more unifunctional commissions, that is, commissions responsible for coordinating single ocean uses. Thus there might be one commission for regulating shared fish stocks, one commission for managing international navigation, one commission for managing renewable energy such as tidal power developments, and another for overseeing offshore oil and gas development.

Each extreme bears positive and negative attributes. On the positive side, a multifunctional approach promises to maximize comprehensive integrated planning and to streamline decisionmaking, since all functions would be handled under a single institutional umbrella. A unifunctional approach, meanwhile, might be more politically attractive since both countries would only have to commit themselves incrementally to the most demanding ocean problems rather than having to take a total plunge into untested, comprehensive ocean management. On the negative side, a multifunctional approach may be politically and practically unrealistic. Given the recent U.S.-Canadian failure to achieve an East Coast fisheries agreement,[40] it may be a political pipedream to expect the countries to agree to an even more complex ocean management program. Given the complexity of modern technologies, the conflicts among interest groups, and the move towards scientific specialization, perhaps it is an illusion to expect a single multifunctional institution to handle the myriad of problems. Perhaps resource management is best left to a negotiation process among unifunctional commissions rather than a centralized decisionmaker. A unifunctional approach, meanwhile, may arguably lead to fragmented, *ad hoc* decisionmaking.

A fourth policy question is what kind of powers should be given to the joint institution(s)? Should they be strictly advisory or should they be granted actual decisionmaking powers? The greater the decisionmaking power, the greater the political resistance is likely to be, for the national executives and bureaucracies are bound to be wary of losing powers to an international body. Political opposition might be curbed, however, by leaving final decisionmaking, with the responsible government agencies in each country. Such might be accomplished by matching bi-national commission membership with responsible agency memberships. For example, if a joint U.S.-Canada Energy Commission were to be established, the Commission might have three members from the Federal Energy Regulatory Commission (the U.S. agency primarily responsible for hydro-electric energy) and three members from the National Energy Board (the Canadian agency responsible for international power exchanges) instead of a membership of independent experts or political appointees.

Fifth, assuming decisionmaking power is granted to the joint institution(s), what type of dispute settlement mechanism should be provided for? Third party arbitration? A Chamber of the International Court of Justice? A special bi-national environment tribunal? Or perhaps no dispute settlement mechanism? In 1979 the American and Canadian Bar Associations recommended to the U.S. and Canadian governments (in addition to a Draft Treaty on Equal Access) a Draft Treaty on a Third-Party Settlement of Disputes which established a potential model for dispute resolution. The Draft Treaty provided for the development (by the exchange of diplomatic notes) of special agreements on third-party settlement for such areas as environmental issues, the management of natural resources of common interest, and pecuniary claims for transboundary damage. In the event no special agreement was reached, disputes would be submitted to a three member arbitral tribunal or a chamber of the International Court of Justice.[41]

The Regional Action Plan Approach

One final institutional arrangement merits consideration. The "action plan" approach to cooperative management of shared marine environments has potential application to the Bay of Fundy - Gulf of Maine region.

Most "action plan" development applied to large scale environmental problems has taken place at the global level. Pehaps the most impressive is the Action Plan prepared for the U.S. Conference on the Human Environment held at Stockholm in 1972 and later endorsed by the U.N. General Assembly.[42]

Equally impressive efforts have been made in recent years to develop action plans at the regional level. The most conspicuous are those that have been negotiated under the auspices of the Regional Seas Program of the United Nations Environment Program (UNEP), an entity created by the U.N. General Assembly in 1973 pursuant to the recommendations of the Stockholm Conference.[43] Regional Action Plans have now been adopted by the appropriate governments in eight designated marine regions: the Mediterranean (1975), the Kuwait Region (1978), the wider Caribbean Region (1981), West and Central African Region (1981), East Asian Region (1981), Red Sea and Gulf of Aden Region (1982), South Pacific Region (1982), and the South-East Pacific Region (1981). Similar plans are being developed for two other regions: the East African Region and the South Asian Seas Region. Action Plan preparation is to be initiated for the South-West Atlantic Region.[44] With the exception of the Mediterranean Action Plan, all all the regional action plans concluded under the Regional Seas Program have focused on regions of ocean space within the developing world.

The principal objective of each plan is set out in the introduction. A typical example of the wording employed is:

> The principal objective of the action plan is the development and protection of the marine environmental and coastal areas of the (appropriate) region particularly for the health and well-being of present and future generations. The action is intended to provide a framework for an environmentally-sound and comprehensive approach to coastal area development particularly appropriate to the needs of the region.

The five elements common to most of the action plans are the sections that deal with environmental assessment, environmental management, legal components, institutional and financial arrangements, and supporting measures.

Since the assessment of the marine environment and adjacent coastal areas must be based on scientific information, agreements to carry out a coordinated regional program of fundamental and applied research is seen as one of the basic activities crucial to the implementation of the other components of the plan. The agreements first articulate this commitment in general terms and then list components that are recommended for inclusion in the coordinated environmental assessment program. Examples of these provisions include: preparation of directories of national research facilities, equipment, manpower, and other resources; training of local scientists and technicians; a compendium of completed, ongoing, and planned research; assessments of oil pollution impacts on the living resources of the region; and cartographic presentation of the sources and concentrations of various pollutants.

The next item common to most of the action plans is the section outlining the preparatory activities regarding environmental management. This component of the action plan framework is based on the philosophy that continuous socio-economic development can best be achieved on a sustainable basis if environmental considerations are taken into account.

Most plans also call for an identification of relevant on-going national, regional, or internationally supported development projects which demonstrate sound environmental management practice. A common component is a commitment to assess present and future environmental impact of development projects, to evaluate the nature and degree of the impact on the marine and coastal environmental, and to identify appropriate means to eliminate or minimize possible damaging effects. Other clauses include commitments to hold regional seminars on coastal area development and coastal zone management and to develop methodologies for the assessment of the impact of projects on the environment.

In addition to these *general* commitments, most plans outline a number of specific projects that the states agree to undertake, such as promotion of the use of environmental impact assessment methodology, determining sites for dumping of hazardous waste, identification of protected areas such as marine parks, wetlands and nurseries, and development of contingency plans for pollution emergencies.

Each of the action plans also contain a legal component which recognizes that the agreements provide a fundamental basis for regional cooperation to protect the marine environment and coastal areas from all environmental hazards, particularly from pollution. To this end, most agreements recognize the importance of ratification and implementation of the relevant international conventions dealing with pollution arising from ship operations, dumping activities, and marine casualties. The legal components may also outline certain activities that are to be undertaken as support measures for the regional convention including: maintenance of an up-to-date compilation of national laws regarding the protection of the marine environment, and, where requested, the provision of technical assistance and advice on drafting appropriate legislation for the effective implementation of the regional convention and other international agreements.

In addition to the action plan, a number of satellite agreements (conventions and protocols) may be part of the package. Generally the conventions and protocols provide further specificity on administrative and operational details. In many cases, for a state to become a contracting party to the main Action Plan convention, it must also agree to sign one or more of the specific protocols.

The Bay of Fundy - Gulf of Maine region differs in many respects from the areas where action plans have been used. The region is bordered by only two states, both of which are highly developed. Collectively and individually the two states have a marine research capability unsurpassed anywhere in the world. Both states have well-developed and entrenched infrastructures to deal with environmental impact assessment and, to varying degrees, coastal zone management. However, many significant differences exist in these infrastructures which make cooperative action extremely difficult.

The question must be asked, is the action plan approach feasible, useful, practical, or desirable as a method of dealing with marine resource and environmental management in the Bay of Fundy - Gulf of Maine region? The answer is a qualified yes. However, if such an approach is to succeed, a great deal of innovative and skillful modification to the existing models would be required.

Achievement of an effective cooperative management and conservation strategy for the region is contingent on three things. First, the arrangement must be politically feasible. Second, it must be adminstratively feasible. Third, it must be environmentally sound. Sound management of the region's environment, although the primary objective of any arrangement, is placed last because the agreement, whatever its form, must first be accepted by both U.S. and Canadian governments. Such an acceptance is a political process that encompasses a wide range of economic and social factors. The U.S. withdrawal of the 1979 East Coast Fisheries Agreement from Senate ratification due to political pressure is only one example of the importance of political acceptability.

The Canada-U.S. arrangement must also lend itself to efficient administration encompassing a broad spectrum of activities, including: coordination of scientific and sociological research, monitoring of activities, evaluation of proposed projects, surveillance and enforcement, and allocation of resources. Any arrangement that does not recognize and take account of the significant differences between U.S. and Canadian approaches to marine resource management may very well not be capable of implementation.

The UNEP Action Plan approach has chosen a blend of approaches ranging from general provisions on the idea of environmental protection to a selection of issue-specific protocols. The general approach usually renders the scheme politically acceptable, but may prove unworkable at the action level because of its non-specificity. Only time will tell.

An Action Plan for the protection and preservation of the marine environment of the Bay of Fundy - Gulf of Maine region is an option worthy of consideration. The settlement of the Bay of Fundy - Gulf of Maine region boundary dispute has, at best, cleared the slate for serious U.S.-Canadian cooperative initiatives. The geographical parameters, activities to be covered, and mechanisms to be employed in an action plan must be closely examined by both countries. Clearly the starting point is a serious commitment, in treaty form, to conserve and manage the region's marine resources. This general commitment, or demonstrative treaty, might also include a package of more specific obligations, one of which should be a mechanism for dealing with energy projects, such as Fundy tidal power, that may have transboundary environmental impacts.

Notes

1. *Environmental Assessment and Review Process Guidelines Order*, SOR/84-467, S. 33(1).

2. For a recent critique of the cost-benefit approach to environmental decisionmaking, *see* T.F. Schrecker, *Political Economy of Environmental Hazards* 39-58 (Law Reform Commission of Canada, Protection of Life Series Study Paper, 1984).

3. For a general discussion of risk analysis and environmental decisionmaking, *see* Gelpe and Tarlock, "The Uses of Scientific Information in Environmental Decision-Making," 48 So. Cal. L. Rev. 371 (1974). For a discussion of the use of risk analysis in the Canadian regulatory context, see Ronson and Swoveland, "Risk Analysis and the Regulation of Safety by Administrative Tribunals: A Case Study," 62 Can. Bar Rev. 4 (1984).

4. Environmental Protection Act, S.N.S. 1973, c. 6, S. 17 (2).

5. *See, e.g.*, *Canada-Nova Scotia Accord for the Protection and Establishment of Environmental Quality.*

6. As an example of the reciprocal legislation approach, Nova Scotia and the federal government entered into a Canada-Nova Scotia Agreement to Govern Offshore Oil and Gas Resource Management and Revenue Sharing on March 2, 1982. The Agreement, which provided for a federal-provincial regulatory board and detailed revenue-sharing, was subsequently implemented through reciprocal legislation. *See* Canada-Nova Scotia Oil and Gas Agreement Act, S.C. 1983-84, C. 29 and Canada-Nova Scotia Oil and Gas Agreement (Nova Scotia) Act, S.N.S. 1984, C.2.

7. P.H. Pearse, F. Bertrand and J.W. MacLaren, *Currents of Change*, Final Report Inquiry on Federal Water Policy 169-170 (September 1985).

8. *See, e.g.*, The TransAlaska Pipeline Authorization Act of 1973, 43 U.S.C. 1651, 1652(b) (1982). The Act exempted pipeline construction and operation from any further action under NEPA beyond the impact statement prepared by the Interior Department in 1972.

9. Personal Communication with FEARO official.

10. *Id.* The Department of External Affairs, in a written submission to the Panel, offered to act as the formal facilitator for interventions by informing the U.S. Department of State and the Danish Ministry of Foreign Affairs of the EARP Panel's willingness to accept foreign interventions and by indicating such interventions would be given in the private capacity of the individuals concerned. Department of External Affairs, *Position Statement on the Beaufort Sea Oil and Gas Proposal* (August 5, 1982).

No Alaskan or Greenlandic interventions were subsequently made at the public sessions, but the Panel did receive written submissions on specific concerns and oral comments from a Greenland resident. Neither the project proponents not the Panel report, however, dealt with the transboundary implications of offshore development in any detail. The Panel merely recommended that the Environmental Assessment Report be made available to the Governments of the United States and Alaska, the North Slope Borough of Alaska, and Government of Denmark and the Home Rule Government of Greenland. Federal Environmental Assessment Review Office, *Beaufort Hydrocarbon Production and Transportation*, Final Report of the Environmental Assessment Panel (July 1984), p. 100.

Recent Cabinet Guidelines assert the need for initiating departments to consider environmental and social effects that are external to Canadian territory. *Environmental Assessment and Review Process Guidelines Order*, SOR/84-467, s. 401(a).

11. Arctic Pilot Project, *Public Interest* 21-23 (Volume 5 of 6 volumes, 1980).

12. New England Governors' Conference, Inc., *Background Paper on the Proposed Bay of Fundy Tidal Power Project* (Prepared for the New England Governors' Conference Inc.) First Annual Bilateral symposium on New England - Eastern Canadian Affairs, May 24-25, 1984, Providence, Rhode Island.

13. Boundary Waters Treaty of 1909, 11 January 1909, United States - Great Britain, 36 Stat. 2448, T.S. No. 548.

14. *Id.*, Article VII.

15. *Id.*, Article VIII.

16. *Id.*, Preliminary Article.

17. *Id.*, Article IV.

18. *Id.*, Article X.

19. *Id.*, Article IX.

20. Cohen, "The Regime of Boundary Waters - The Canadian-United States Experience," 146 Recueil des Cours 217, 289 n. 164 (1975 III).

21. *E.g.*, the current (1978) Great Lakes Water Quality Agreement, 30 U.S.T. 1383, T.I.A.S. No. 9257, was developed from the 1972 Agreement, 23 U.S.T. 302, T.I.A.S. No. 7312, which itself grew out of a series of references on Great Lakes water quality.

22. International Joint Commission, "Investigation of the International Passamaquoddy Tidal Power Project" (1961) [IJC Docket 72].

23. For an overview of several of those suggestions, *see* Munton, "Paradoxes and Prospects," in *The International Joint Commission: Seventy Years On* 64-78 (R. Spencer, J. Kirton, K. Nossal, eds., 1981).

24. *See e.g., id.* at 74-75.

25. The agreements are discussed in O.P. Dwivedi and J.E. Carroll, "Issues in Canadian - American Environmental Relations" in O.P. Dwivedi (ed.) *Resources and the Environment: Policy Perspectives for Canada* 306, 315-317 (1980). For a discussion of the importance of provincial inputs into bi-national cooperation, *see* J. Piette, "The Role of Provincial Governments in the Field of Transfrontier Pollution," in *Common Boundary/Common Problems: The Environmental Consequences of Energy Production* 69-70 Proceedings of a Conference held at Banff, Alberta, Canada, March 19-21, 1981 (1982).

26. For a summary of the discussions and actions of the first eleven Conferences, *see New England Governors' and Eastern Canadian Premiers' Conference Eleven-Year Overview of Discussions* (1984).

27. *Resolutions*, Twelfth Annual Conference of the New England Governors and the Eastern Canadian Premiers, Newport, Rhode Island (June 17-19, 1984).

28. "Leaders to Help Regulatory Process," (Halifax) *Chronicle Herald*, p. 5, June 19, 1985.

29. *Record of Discussion 13-11, Bay of Fundy Tidal Power*, Thirteenth Annual Conference of the New England Governors and Eastern Canadian Premiers, St. Andrews, N.B. (June 16-18, 1985).

30. Transport Canada, Coast Guard and U.S. Department of Transportation, Coast Guard, *Canada-United States Joint Marine Pollution Contingency Plan* (September 15, 1983).

31. Personal Communication, Dr. Adam Kerr, Canadian Hydrographic Service (January 29, 1986).

32. For a general discussion of the major types of international agreements - demonstrative, administrative, distributive, and resolutive, *see* D. Johnston (ed.), *The Environmental Law of the Sea* 356-357 (1981). The types of agreements are not necessarily mutually exclusive, that is, a single treaty might include demonstrative, administrative, distributive, and resolutive elements.

33. The Agreement is reprinted in 23 Int'l Legal Materials 269 (1984).

34. American Bar Association and Canadian Bar Association, *Settlement of International Disputes Between Canada and the U.S.A.* (1979). The Draft Treaty on Equal Access and Remedy in Cases of Transfrontier Pollution is also reprinted in Comment, "Who'll Stop the Rain: Resolution Mechanisms for U.S.-Canadian Transboundary Pollution Disputes," 12 Den. J. Int'l L.&P. 51, Appendix A (1982). For a discussion of the Joint Working Group's work, *see* G. Alexandrowicz, "A Proposal to Assist the Resolution of Environmental Disputes Between Canada and the United States," in *Common Boundary/Common Problems: The Environmental Consequences of Energy Production* 58-60, Proceedings of a Conference held at Banff, Alberta, Canada, March 19-21, 1981 (1982).

35. For the text and commentary on the model act, *see* Uniform Law Commission Proceedings 1982, Appendix DD, p. 500.

36. As of October 1984 Montana, Colorado, and New Jersey were members of the "club", and the Act had been introduced into the Wisconsin and Minnesota legislatures. "The Trickle Down Theory of International Law," *Canadian Council on International Law Bulletin*, Vol. XI, No. 2 (February 1985), p.4. On June 10, 1985 the Ontario legislature gave first reading to the Transboundary Pollution Reciprocal Access Act as Bill 3 of the 1st session of the 33rd Legislature. Personal Communication, Mr. Doug Beecroft, Ministry of the Attorney General of Ontario (January 29, 1985).

37. S.C. 1970-71-72, c. 47.

38. *Id.*, S. 21.1(l) *as amended by* S.C. 1980, c. 45, s.3.

39. *Id.*, S. 21.1(2)(b) *as amended by* S.C. 1980, c. 45, s.3.

40. For a discussion of the numerous factors behind the agreement's demise, *see* D. VanderZwaag, *The Fish Feud: The U.S. and Canadian Boundary Dispute* 90-92 (1983).

41. The text of the Draft Treaty on a Third-Party Settlement of Disputes is reprinted in Comment, "Who'll Stop the Rain: Resolution Mechanisms for U.S.-Canadian Transboundary Pollution Disputes," 12 Den. J. Int'l L.&P. 51, Appendix B (1982).

42. Report of the U.N. Conference on the Human Environment, U.N. Doc. A/Conf.48/14.

43. In December 1982 the U.N. General Assembly adopted 11 resolutions concerning various environmental matters including the establishment of the United Nations Environmental Programme. *See* (1973) 10 U.N. Monthly Chronicle 72-73. *See also* Yearbook of the United Nations 1972, pp. 323-329.

44. For a recent overview of the status of the Regional Action Plan, *see* Neuman, "The United Nations Regional Seas Program," *Marine Tehnology Society Journal*, Vol. 19 No. 1 (1985), pp. 46-52.

Chapter 7

Summary of Conclusions and Recommendations

The research described in this monograph focused on four principal questions:

1. What are the potential transboundary effects of the proposed Fundy tidal power project and how adequate is the scientific information underlying these predictions?

2. What is the nature and adequacy of the legal and institutional frameworks for environmental decisionmaking in Canada and in the United States?

3. How well is scientific information incorporated into the environmental assessment practices of Canada and the U.S., particularly in regard to indirect and transboundary effects?

4. What institutional changes may be appropriate to improve transboundary environmental decisionmaking?

This chapter summarizes the major findings of the preceding chapters and sets forth several recommendations for future work. The findings and recommendations are presented in the two major areas of investigation: scientific questions and legal and governmental considerations.

Scientific Research Needs for Fundy Tidal Power

The Bay of Fundy - Gulf of Maine ecosystem exhibits broad seasonal, annual, and other periodic variations. This dynamic character makes it difficult to predict far-field, transboundary effects. Effects which have been associated with the tidal power project include increases in tidal range along the Canadian and New England coasts, changes in water column mixing, and disruptions in the routes of migratory species. As all of the postulated effects, with one exception, are only changes in magnitude of existing ecosystem processes, the question arises whether these incremental changes will exceed normal system variation, i.e., whether the changes will be detectable, measurable, and attributable to the Fundy tidal power project. The exception concerns the shad fishery. Research suggests that a large proportion of the East Coast's migratory shad use the Upper Bay of Fundy as a summer feeding ground and that they are likely to experience high mortality as they pass through the turbines, while following the tidal stream.

Most of the predicted effects on the physical oceanography which would engender biological effects derive from a computer model of the Bay of Fundy - Gulf of Maine system prepared by Dr. David Greenberg of the Bedford Institute of Oceanography in Dartmouth, Nova Scotia. His work builds on a twenty-year effort to understand the factors behind the high tides of the Bay of Fundy. Dr. Greenberg's work demonstrates that the natural oscillatory period of the tides formed by the geography of the Maine-Fundy basin corresponds closely to the lunar tidal period, thus creating resonant effects. The Greenberg model predicts that the shortening of the basin by the construction of a tidal barrage at the proposed site (Minas Basin-Economy Point) would bring the period into even closer resonance, increasing the tidal amplitude in the lower Bay of Fundy and Gulf of Maine by approximately 6-9 inches, i.e., high tides would be 6-9 inches higher and low tides 6-9 inches lower.

Accuracy of the Greenberg Model

The Greenberg model predictions of the physical oceanography are very good, differing from observed values by less than 10%. Because of its accuracy, it would not be worthwhile to redo the model; however, work should proceed on fine-tuning it in three areas: 1) resolution of depths in areas of rough bottom topography (the model uses an average depth); 2) improving the resolution in deeply indented coastal areas for more accurate determination of local impacts by coupling the Greenberg model with existing coastal models in areas of concern; and 3) coupling the model with more refined energy dissipation models for greater accuracy.

Mean Sea Level and Resonance

Mean sea level has been rising along the northeast coast of North America for the last 8000 years, and this process may be accelerating because of the "greenhouse effect." These increases affect the geometry of the ocean basin, thus potentially altering the natural oscillatory period. Depending on basin geometry and current dissipation, increased sea level may push the Bay - Gulf system away from resonance, countering the effects of the Fundy tidal power barrage. Basin geometry and energy dissipation need further investigation. (See recommendation –1 above.) As some coastal areas have higher rates of submergence than others, more data on submergence should be gathered to pinpoint these more sensitive areas.

Coastal Impacts

Impacts of higher tides and storm surges will be most noticeable in those areas that now suffer erosion from storms, notably those modified and stabilized by human activities. Research shows that most mid-coast Maine beaches are accretionary and should not suffer from increased tidal amplitude, except perhaps in the rare case when storm surges coincide with the highest high tides. However, beaches further south and those modified by human structures are at risk. Research should continue on the relationship between erosion and beach stabilization.

Mixing Processes

Shallow water mixing processes play a critical role in the biological productivity of the region. Tidal power may enhance this mixing, heightening general productivity, but whether this will benefit the ecosystem is unclear, because there is much we do not know about the adaptation of organisms to the existing conditions and their responses to change. Research should continue on the physical-biological oceanographic link. However, because of the complexity of the issues, research directed specifically at measuring the effects of tidal power on mixing will probably yield little concrete evidence.

Migratory Species

In addition to shad, preliminary research suggests that other fish species such as bass, gaspereau, and dogfish may use the upper Bay of Fundy as a feeding ground, while certain shorebirds may rely heavily on intertidal feeding areas that will be affected by the dam. Research should continue on the identified fish and fowl feeding in the area to determine more about their abundances and habits, so that better predictions can be made on the likely impacts on these species and possible mitigation measures can be explored.

General

The limitations on our abilities to detect, measure, and attribute specific effects to tidal power mean that decisions on research priorities and on proceeding with the project itself must ultimately be based on judgment. Long term, cooperative (United States-Canada) study of the dynamics of the Bay - Gulf system should continue.

Law and Governmental Institutions

Difficulties between the U.S. and Canada in achieving joint or cooperative management of resources in the Bay of Fundy - Gulf of Maine system are evident in the referral of the Gulf of Maine boundary delimitation to the International Court of Justice, the breakdown of discussions concerning allocation and management of the Georges Bank fish stocks, and the conflict over oil tanker passage through Head Harbor Passage to the once-proposed Pittston oil refinery in Eastport, Maine. Domestic and international law and bilateral arrangements all have a role to play in the resolution of transboundary issues in this region.

International Law

Customary international law recognizes the sovereign right of each nation to use its resources as it sees fit, but imposes upon each nation the obligations not to use or permit the use of its territory in a manner that causes significant environmental damage to another nation, and to pay compensation if transboundary environmental damage does occur. The United States and Canada recognize these obligations.

The United States and Canada, in both their practice and their stated support for principles promoted by the United Nations and the Organization for Economic Cooperation and Development (OECD), appear to recognize a procedural obligation of notice and consultation prior to commencing activities having significant transboundary effects, but as notice and consultation have varied from situation to situation, the extent of the "obligation" is uncertain.

The U.S. President and State Department, and the Canadian Prime Minister and Office of External Affairs, should take the opportunity presented by Fundy tidal power to make explicit that both countries recognize an international law obligation to notify and consult on projects with potential transboundary impacts and to specify procedures to be followed in such cases.

Domestic Law

United States. While the export of power requires a permit with review for compliance with standards, there is no analogous regulation of power imports and accompanying environmental review. Any potential review would be via the Presidential Permit process for transboundary transmission lines, administered by the Department of Energy (DOE), and preparation of an Environmental Impact Statement, but whether such a statement would examine transboundary effects of a foreign facility over which DOE has no jurisdiction is uncertain. Nor does state regulatory review appear to provide an avenue for consideration of such effects.

Failure to exercise controls over energy import activities that could result in forseeable environmental damage, could, under developing theories of international law, result in the United States being held to have consented to the production of such damage and the waiving of the right to compensation.

U.S. environmental review procedures do not necessarily incorporate consideration of impacts in a foreign country.

The United States should consider legislation requiring a license for the importation of power and requiring consideration of environmental implications of any such importation.

The Council on Environmental Quality (CEQ) should amend its National Environmental Policy Act (NEPA) regulations to define "transboundary regions" and to require full consideration of effects of federal actions on both sides of any international boundary. To facilitate this consideration, the regulations should authorize Canadian public participation in EIS preparation on a par with U.S. public participation and specify procedures for intergovernmental consultation when transboundary regions are affected.

Canada. The current Environmental Assessment Review Process (EARP) is more discretionary than the NEPA process, which, while allowing flexibility, creates uncertainty regarding (1) which projects are subject to a review (e.g., participation by crown corporations such as Tidal Power Corporation and Nova Scotia Power Corporation is voluntary); (2) the scope of review; (3) whether and to what extent transboundary effects will be considered; and (4) the degree to which panel recommendations are reinforced through project monitoring.

There is some duplication and lack of coordination of environmental review, particularly at the federal level between the National Energy Board (NEB) and the Environmental Assessment Review Process. (Coordination of the federal-provincial reviews is accomplished through environmental accords between the provinces (e.g., Nova Scotia and New Brunswick) and the Federal Department of Environment.)

The Canadian Government should establish permanent guidelines or regulations covering such factors as notice of hearings, time tables, evidentiary rules, the "scoping" process, and the requirement of post-project monitoring.

The Government should delegate to the Minister of the Environment authority to "internationalize" environmental reviews involving transboundary regions, providing for participation by the U.S. government and citizens on a reciprocal basis with U.S. procedures for Canadian participation and for joint scientific research committees.

EARP and NEB environmental reviews should be coordinated to avoid duplication and to provide for joint hearings or for cross attendance and participation at related hearings on the same project.

There currently exists a bilateral U.S.- Canadian body, the International Joint Commission (IJC), established by bilateral convention, the Boundary Waters Treaty Act of 1909, which has the authority and demonstrated capacity to study transboundary impacts of complex project proposals and make recommendations on mitigation measures when both countries agree to refer a problem to the Commission. The IJC previously investigated tidal power under a joint reference on the proposed Passamaquoddy tidal power project in the 1960s.

The U.S. and Canadian governments should enter into negotiations toward the submission to the IJC of a reference on the transboundary impacts of the Fundy tidal power project. Once the reference is made, the reports of the IJC study board, along with the final recommendations, should be incorporated into the EIS under the United States Presidential Permit process and the environmental assessment prepared under the Canadian EARP.

Appendices

A. Canada - U.S. Energy Exchange Workshop, October 24, 1983, Portland, Me.: Agenda and List of Participants.

B. Workshop on Regional Scientific Needs on Fundy Tidal Power, November 20-21, 1983, Durham, N.H.: Agenda and List of Particpants.

C. List of Project Working Papers.

APPENDIX A

CANADA-UNITED STATES ENERGY EXCHANGE WORKSHOP

October 24, 1983
University of Maine School of Law, Portland, Maine

Workshop Agenda

Sponsored by: Marine Law Institute
Northeast-Midwest Institute
Canadian-American Center, University of Maine

9:00 - 9:15	Welcome and Introductions
9:15 - 10:00	**Garet A. Bornstein, U.S. Department of Energy:** Existing Regulatory Framework for: Electricity Exports, Presidential Permits, Natural Gas Imports/Exports
10:00 - 10:45	**Thomas Keating, Carleton University,** Existing Regulatory Framework for Electricity Imports and Exports in Canada
10:45 - 11:00	Coffee break
11:00 - 11:45	**Diane DeVaul, Northeast-Midwest Institute,** Potential for Canada-United States Electricity Trade
11:45 - 1:00	Lunch
1:00 - 2:30	Fundy Tidal Power as a Source of Electricity for the United States 1. **Allison Rieser, MLI** and **David VanderZwaag, Dalhousie Ocean Studies Programme,** overview of Case Study, Fundy Tidal Power 2. **Karen Massey, MLI,** Discussion of Electricity Exchanges Policy and Legislative Proposal for United States Regulation of Electric Power Imports
2:30 - 4:00	**Round Table Discussion** (Attendees): Discussion of Canada-United States energy policy, legislative proposal, other legislative and institutional options for dealing with energy exchanges and transboundary issues (e.g., International Joint Commission)
4:00	Reception

LIST OF ATTENDEES

Speakers:

Garet A. Bornstein
Office of Fuels Programs
Economic Regulatory Administration
Department of Energy
Room GA-017
1000 Independence Avenue, S.W.
Washington, D.C. 20585

Diane DeVaul
Senior Policy Analyst
Northeast-Midwest Institute
218 D Street, S.E.
Washington, D.C. 20003

Thomas Keating
Professor
Department of Political Science
Loeb Building
Carleton University
Ottawa, Ontario
CANADA K1S 5B6

Karen A. Massey
Staff Attorney
Marine Law Institute
246 Deering Avenue
Portland, Maine 04102

Alison Rieser
Director
Marine Law Institute
246 Deering Avenue
Portland, Maine 04102

David VanderZwaag
Staff Attorney
Dalhousie Ocean Studies Programme
1321 Edward Street
Halifax, Nova Scotia
CANADA B3H 3H5

Discussants:

Ingrid Bartinque
Consultant
Massachusetts Coastal Zone Program

David Lamoine
Senior Field Representative
Office of Senator George S. Mitchell
151 Forest Avenue
Portland, Maine 04101

Donat B. Boisvert
Director
Maine Canadian Legis. Advisory
State House Station 107
Augusta, Maine 04333

Mary Cerullo
Marine Communicator
NH/Maine Sea Grant College Program
University of Maine
Orono, Maine 04469

Steve Colman
Coastal Zone Liason
New England Governor's Conference
76 Summer Street
Boston, Massachusetts 02110

Edward Collins
Professor
Canadian-American Center
154 College Avenue
Orono, Maine 04469

Susan Morrell
Energy Law Institute
Franklin Pierce Law Center
2 White
Concord, New Hampshire 03301

Bruce Ellsworth
Chief Engineer
N.H. Utilities Commission
8 Old Suncook Road
Concord, New Hampshire 03301

Carl Flynn
Manager, Strategic Development
New Brunswick Electric Power Commn.
P.O. Box 2000
Fredricton, N.B.
CANADA E3B 4X1

Mary Hain
Energy Law Institute
Franklin Pierce Law Center
2 White
Concord, New Hampshire 03301

Constance Irland
Deputy Director of Planning and Conservation
Office of Energy Resources
State House Station 53
Augusta, Maine 04333

Arthur Johnson
New Hampshire Public Utilities Commission
8 Old Suncook Road
Concord, New Hampshire 03301

David Keeley
Supervisor, Natural Resource Programs
State Planning Office
184 State Street
Augusta, Maine 04333

Kim Kenway, Esq.
Staff Attorney
Maine Public Utilities Commission
242 State Street
Augusta, Maine 04333

Ms. Constance LaPointe
Director
Energy Program
New England Governor's Conference
76 Summer Street
Boston, Massachusetts 02110

Faye Luppi
Intern
Marine Law Institute
246 Deering Avenue
Portland, Maine 04102

Marc St. Laurent
Canadian Consulate
500 Boylston Street
Boston, Massachusetts 02116

Dr. Judith Spiller
Professor of Research
Complex Systems Research Center
University of New Hampshire
O'Kane House
Durham, New Hampshire 03824

Allen L. Springer
Professor
Department of Government
Bowdoin College
Burnswick, Maine 04011

Robert J. Sugarman, Esq.
Sugarman & Denworth
Suite 510, North American Bldg.
121 South Broad Street
Philadelphia, Pennsylvania 19107

Robert Taylor
Staff Attorney
Marine Law Institute
246 Deering Avenue
Portland, Maine 04102

APPENDIX B

WORKSHOP ON REGIONAL SCIENTIFIC NEEDS IN FUNDY TIDAL POWER DEVELOPMENT, November 20-21, 1983, Durham, N.H.

AGENDA

Sunday, November 20

Dinner - Wildwood Lounge, New England Center	6:00 PM

SESSION I - Hetzl Room, Alumni Center

A. Introduction 1. Bell-Donner Project 2. Purpose of Workshop 3. Utilization of Results	7:45 PM
B. Fundy Tidal Power Background 1. Fundy Tidal Power Project 2. Fundy Environmental Studies Committee a. Studies b. Results c. Future research	8:00 PM
C. Regional Effects 1. Categories of Effects 2. Sources of Information 3. Bigelow Study	8:45 PM
D. Agenda Discussion	9:15 PM

Monday, November 21

SESSION II - 1925 Room, Alumni Center

A. Greenberg Model	8:30 AM

1. Summary
2. Prediction
   a. Amplitude and phase changes
   b. Effect on tidal range, currents, areas of mixing, and surface water temperature

3. Relationship between tidal regime and mean circulation in Bay of Fundy/Gulf of Maine

4. Factors modifying tides
   a. Meteorology (storms, winds)
   b. Coastal and basin configuration
   c. Astronomical effects

5. Accuracy of model and prediction refinement
   a. Open boundary problem

b. Depth averaging of bottom waters
c. Comprehensive tidal data

6. Alternate models

B. Coastal Effects - Shoreline Modification 9:30 AM
1. Regional sea level change
a. Rate
b. Uniformity

2. Sedimentary processes
a. Beaches
1. equilibrium state vs erosion-accretion
2. effect of altered tidal range and currents
3. storms and storm surges
4. uniformity of regional response

b. Embayments
1. channel, inlets and shoals and near shore currents
2. effect of altered tidal range and currents

c. Measurement techniques
1. baseline surveys
2. simulations
3. synoptic surveys
4. process studies

3. Intertidal biota
a. Faunal response
b. Displacement and recovery times
c. Measurement techniques

Coffee Break 10:30

4. Summary 10:45

C. Coastal Effects - Estuaries 11:00 AM

1. Estuarine mixing
a. Resonance effects
b. Tidal prism
c. Tidal and residual circulation
d. Flushing rates
e. Nutrient fluxes
f. Bottom topography and tidal level
g. Estuarine/shelf water exchange processes
h. "Typical" northeast estuary?
i. Effect of tidal range and current changes on estuarine processes
1. types of effects
2. measurement of effects
3. recovery times

2. Estuarine productivity
a. Salt marsh
b. Mud flats
c. Water column
d. Benthos

e. Effect of tidal range and current changes on productivity
   1. types of effects
   2. measurement
   3. recovery times

3. Summary 12:30 PM

Lunch 12:45 PM

Monday, November 21

SESSION III - 1925 Room, Alumni Center

A. Mixing 1:45 PM
   1. Hydrodynamics
      a. Vertical and horizontal mixing
      b. Mixing and water temperature
      c. Productivity and mixing
      d. Seasonal and annual influences on mixing

   2. Expanded mixing
      a. Geographic areas - southwest Nova Scotia and Great South Channel
      b. Surface water temperature
      c. Effects on migratory and resident species

   3. Measurement of change

Coffee Break 3:25 PM

   4. Summary 4:00 PM

B. Migratory species 4:30 PM
   1. Shad
      a. Utility of fishways
      b. Adaptation, recovery times

   2. Shorebirds
      a. Crowding
      b. Adaptation, recovery times

   3. Summary

Adjourn 5:30 PM

List of Attendees

Participants:

Dr. Wendell Brown
Department of Earth Science
University of New Hampshire
Durham, New Hampshire 03824

Dr. Ken Drinkwater
Bedford Institute of Oceanography
Dartmouth, Nova Scotia
Canada B2Y 4A2

Dr. Ken Fink
Ira C. Darling Center
University of Maine
Walpole, ME 04573

Dr. Don Gordon
Bedford Institute of Oceanography
Dartmouth, Nova Scotia
Canada B2Y 4A2

Mr. David Keeley
State Planning Office
State of Maine
184 State Street
Augusta, ME 04016

Dr. Bryan Pearce
Civil Engineering Department
455 Albert Hall
University of Maine
Orono, ME 04473

Dr. Peter Ricketts
Department of Geography
St. Mary's University
Halifax, Nova Scotia
Canada B3H 3C3

Dr. David Scarratt
Department of Fisheries and Oceans
Biological Station
St. Andrews, New Brunswick
Canada E0G 2X0

Dr. Michael Sissenwine
National Marine Fisheries Service
Northeast Center
Woods Hole, MA 02543

Dr. Jerry Topinka
Bigelow Laboratory
West Boothbay Harbor, ME 04575

Dr. Les Watling
Ira C. Darling Center
University of Maine
Walpole, ME 04573

Sponsors:

Dr. Arthur Hanson
Institute for Resource and Environmental Studies
Dalhousie University
6086 University Avenue
Halifax, Nova Scotia
Canada B3H 1W7

Dr. Judith Spiller
Complex Systems Research Center
University of New Hampshire
Durham, NH 03824

Mr. John Roanowicz
Complex Systems Research Center
Univeristy of New Hampshire
Durham, NH 03824

APPENDIX C

INTERNATIONAL RISKS OF COASTAL ENERGY DEVELOPMENT IN THE GULF OF MAINE - BAY OF FUNDY REGION

List of Project Working Papers

1. The Proposed Pittston Oil Refinery: A Case Study of U.S. Environmental Decisionmaking

2. Overview of the Adequacy and Analysis of Scientific Information Concerning the Transboundary Effects of Fundy Tidal Power

3. Regional Scientific Needs in Fundy Tidal Power Development - A Report of a Workshop

4. International Law and Prosed Energy Developments in the Bay of Fundy - Gulf of Maine Region: Liability for Transboundary Pollution Damage and State Responsibility

5. Private Remedies for Transboundary Environmental Injury Due to Coastal Energy Development in the Bay of Fundy - Gulf of Maine Region

6. The International Joint Commission and Transboundary Environmental Impacts in the Bay of Fundy - Gulf of Maine Region

7. United States Regulation of Dometic Tidal Power Development

8. Existing United States Regulatory Authority Applicable to Nova Scotian Power Development

9. Fundy Tidal Power: A Technology Assessment System Case Study

10. Canadian Law Relating to Tidal Power Development and Oil Terminal Siting and Oil Tankering

11. The Characteristics and Role of Scientific Information in the Canadian Environmental Assessment and Review Process

12. Regional Action Plans for the Protection of the Marine Environment: Model for the Bay of Fundy - Gulf of Maine?

SUBJECT INDEX

Note: "Passim" means that the topic is referred to througout the indicated chapter. Canadian governmental agencies are listed under headings beginning with "Department," while the U.S. governmental agencies begin under "U.S."